Converged Network Service Architecture

Electronic Business

Herausgegeben von Christine Strauss

Band 2

PETER LANG
Frankfurt am Main · Berlin · Bern · Bruxelles · New York · Oxford · Wien

Natalia Kryvinska

Converged Network Service Architecture

A Platform for Integrated Services Delivery and Interworking

PETER LANG
Internationaler Verlag der Wissenschaften

Bibliographic Information published by the Deutsche Nationalbibliothek
The Deutsche Nationalbibliothek lists this publication in the Deutsche Nationalbibliografie; detailed bibliographic data is available in the internet at http://dnb.d-nb.de.

Cover design:
Olaf Glöckler, Atelier Platen, Friedberg

Printed with the support of the Federal Ministry of Science and Research in Vienna.

ISSN 1868-646X
ISBN 978-3-631-59525-1

www.peterlang.de

"If we knew what it was we were doing, it would not be called research, would it?"

Albert Einstein

(http://www.brainyquote.com/quotes/quotes/a/alberteins148837.html, last accessed 28.11.09)

Table of Content

List of Figures

List of Tables

List of Acronyms

ACK	Acknowledgement
AcQ	All call Query
ADSL	Asymmetric Digital Subscriber Line
AGW	Access Network Gateway
ANSI	American National Standards Institute
API	Application Programming Interface
ASN.1	Abstract Syntax Notation No. 1
ASP	Application Service Providers
ATM	Asynchronous Transfer Mode
BAS	Broadband Access Server
BCSM	Basic Call State Model
BGP	Border Gateway Protocol
B-ISDN	Broadband ISDN
C2D	Click-to-Dial
CAMEL	Customized Application of Mobile Enhanced Logic
CAS	Channel-Associated Signaling
CC	Call Control
CCAF	Call Control Agent Function
CCBS	Completion of Call to Busy Subscriber
CCF	Call Control Function
CCIB	Call Completion Internet Busy
CCITT	International Telegraph and Telephone Consultative Committee

CCS	Common Channel Signaling
CG	Customer Gateway
CLI	Calling Line Identification
COM	Component Object Module
COPS	Common Open Policy Service
CORBA	Common Object Request Broker Architecture
CP	Control Platform
CPE	Customer Premises Equipment
CPL	Call-Processing Language
CPU	Central Processor Unit
CS 1/2/3	Capability Set 1/2/3
C-SIP	Core SIP network server
CSM	Communication Session Manager
CSP	Communication Service Provider
CT	Computer Telephony
CTI	Computer Telephony Integration
DB	Database
DCE	Distributed Computing Environment
DCOM	Distributed Common Object model
DECT	Digital Enhanced Cordless Telecommunications
DHCP	Dynamic Host Configuration Protocol
DiffServ	Differentiated Services
DP	Detection Point
DPE	Distributed Processing Environment
DSL	Digital Subscriber Line

DSLAM	Digital Subscriber Line Access Multiplexer
DSM-CC	Digital Storage Multimedia Command and Control protocol
DSP	Digital Signal Processor
DSS1	Digital Subscriber Signaling System no. 1
DSS2	Digital Subscriber Signaling System no. 2
DTMF	Dual-Tone MultiFrequency
DWDM	Dense Wavelength Division Multiplexing
ECA	Electronic Components, Assemblies, & Materials Association
ECMA	European Computer Manufacturers Association
ECTF	Enterprise Computer Telephony Forum
ECTRA	European Committee on Telecommunications Regulatory Affairs
EGK	Extended Gate Keeper
ENUM	E.164 Number Mapping
ESP	External Service Provider
ETO	European Telecommunications Office
ETSI	European Telecommunications Standards Institute
EU	Europian Union
FIFO	First-In-First-Out
FMC	Fixed-Mobile Convergence
FPLMTS	Future Public Land Mobile Telecommunication System
GDB	Gateway DataBase
GGSN	Gateway GPRS Support Node
GII	Global Information Infrastructure

GK	Gatekeeper
GSM	Global System for Mobile Communications
GTT	Global Title Translation
GUI	Graphical User Interface
GW	Gateway
HSS	Home Subscriber Server
HTML	HyperText Mark-up Language
HTTP	HyperText Transfer Protocol
ICW	Internet Call Waiting
ID	Identifier
IDL	Interface Definition Language
IETF	Internet Engineering Task Force
IIOP	Internet Inter-Orb Protocol
ILEC	Incumbent Local Exchange Carrier
IMTC	International Multimedia Teleconferencing Consortium
IN	Intelligence Network
INAP	IN Application Protocol
INCM	IN Conceptual Model
IntServ	Integrated Services
IOP	Inter-Object request broker Protocol
IOR	Interoperable Object Reference
IP	Internet Protocol
IP	Intelligent Peripheral
IPC	InterProcess Communication
IRTF	Internet Research Task Force

IS41-Cb	Interim Standard 41, revision C
ISC	International Softswitch Consortium
ISDN	Integrated Services Digital Network
ISP	Internet Service Provider
ISUP	ISDN Signaling User Part
IT	Information Technology
ITSP	Internet Telephony Service Provider
ITU	International Telecommunication Union
ITU-T	International Telecommunications Union
IUAP	Interworking User Application
IVR	Interactive Voice Response
IWF	InterWorking Function
IWU	Interworking Unit
JAIN	Java Advanced Intelligent Network
JCC	Java Call Control
JDMK	Java Dynamic Management Kit
JTAPI	Java Telephony API
LAN	Local Area Network
LDAP	Lightweight Directory Access Protocol
LDMS	Local Multipoint Distribution Service
LNP	Local Number Portability
LX	Local Exchange
M3UA	MTP layer 3 User Adaptation layer
MAP	Mobile Application Part
MEGACO	IETF Media Gateway Control Protocol

MGC	Media Gateway Controller
MGCP	Media Gateway Control Protocol
MGW	Media Gateway
MIB	Management Information Base
MPLS	Multi-Protocol Label Switching
MS	Meeting Scheduler
MSC	Mobile Switching Center
MTP	Message Transfer Part
MTP L1/2/3	Message Transfer Part Level 1/2/3
NAS	Network Access Server
NAT/PAT	Network/Port Address Translation
NE	Network Element
N-ISDN	Narrowband ISDN
NGN	Next-Generation Network
NP	Number Portability
NSP	Network Service Provider
OAM	Operation, Administration and Maintenance
O-BCSM	Originating BCSM
OLTP	On-line Transaction Processing
OMAP	Operations, Maintenance and Administration Part
OMG	Object Management Group
ORB	Object Request Broker
OSA	Open Service Architecture
OSI	Open Systems Interconnection

O-SIP	Originating SIP network server
OSS	Operating Support System
OTM	Object Transaction Manager
PBN	Packet Based Network
PBX	Private Branch Exchange
PC	Personal Computer
PCM	Pulse Code Modulation
PDA	Personal Digital Assistant
PDN	Packet Data Network
PIC	Point In Call
PIN	Personal Identification Number
PINT	PSTN Internet Interworking
PLMN	Public Land Mobile Network
PNC	Public Network Computing
POP	Point of Presence
POTS	Plain Old Telephony System
PPP	Point-to-Point Protocol
PSTN	Public Switched Telephone Network
QoS	Quality-of-Service
RADIUS	Remote Authentication Dial-In User Service
RAS	Registration Admission and Status
RCP	Real-time Control Protocol
RFC	Request For Comment
RFI	Request For Information
RG	Residential Gateway

RGW	Residential gateway
RMI	Remote Method Invocation for Java
RP	Reference Point
RSVP	Resource Reservation Protocol
RTP	Real Time Protocol
SCCP	Signaling Connection Control Part
SCE	Service Creation Environment
SCF	Service Control Function
SCN	Switched Circuit Network
SCP	Service Control Point
SCTP	Stream Control Transmission Protocol
SDF	Service Data Function
SDP	Session Description Protocol
SG	Signaling Gateway
SGCP	Simple Gateway Control Protocol
SIB	Service Independent Building Block
SIP	Session Initiation Protocol
SLA	Service Level Agreement
SMS	Service Management System
SMS	Short Messages System
SMTP	Simple Mail Transfer Protocol
SN	Serving Node
SNMP	Simple Network Management Protocol
SP	Service Provider
SP	Signaling Point

SPAN	Services and Protocols for Advanced Networks
SPC	Signaling Point Code
SPC	Stored Program Control
SPIRITS	PSTN/IN Requesting Internet Service
SS7	Signaling System number 7
SS8	Signaling System No.8
SSF	Service Switching Function
SSL	Secure Sockets Layer
SSP	Service Switching Point
STP	Signaling Transfer Point
TA	Travel Agency
TAPI	Telephony Application Protocol Interface
T-BCSM	Terminating BCSM
TC	Transaction Capabilities
TCAP	Transaction Capability Application Part
TCP	Transmission Control Protocol
TDM	Time Division Multiplex
TDP	Trigger Detection Point
TGW	Trunk gateway
TINA	Telecommunications Information Networking Architecture
TIPHON	Telecommunications and Internet Protocol Harmonization Over Networks
TMN	Telecommunication Management Network
TP	Transport Protocol
TRIP	Telephony Routing over Internet Protocol

T-SIP	Terminating SIP network server
TUP	Telephone User Part
TX	Transit Exchange
UDP	User Datagram Protocol
UML	Unified Modeling Language
UMTS	Universal Mobile Telecommunications System
UPT	Universal Personal Telecommunication
VAR	Value-Added Reseller
VoIP	Voice over IP
VP	Virtual Presence
VPN	Virtual Private Network
VSL	Virtual Second Line
WAN	Wide Area Network
WAP	Wireless Application Protocol
WG	Working Group

List of Mathematical Symbols

λ	Average arrival rate to system
μ	Average service rate
ρ	Server utilization
σ	Standard deviation
α	Positive real root
θ	Probability that arriving customer finds server busy
$\lambda(j)$	Average arrival rate to system for jth priority class
σ_g^2	Variance of duration of busy period
σ_L^2	Variance of number of users in system
σ_{LW}^2	Variance of number of users waiting
σ_{LA}^2	Variance of number of users in system at user arrival instants
σ_T^2	Variance of time in system
σ_{TA}^2	Variance of inter-arrival time
σ_{TS}^2	Variance of service time
$\sigma_{Ts(j)}^2$	Variance of service time for jth priority class
σ_{Tw}^2	Variance of waiting time
λA	Actual arrival rate to servers
C_A^2	Squared coefficient of variation of inter-arrival time
C_S^2	Squared coefficient of variation of service time
$C_{S(j)}^2$	Squared coefficient of variation of service time for jth priority class
D	Deterministic distribution
E_k	k-stage Erlangian distribution

$f*(s)$	Laplace-Stieltjes transform of pdf of inter-arrival times
$f_{Si}(t)$	Service time probability distribution function
G	General distribution
g_1	Average duration of busy period
H_k	k-stage Hyperexponential distribution
K	Size of finite storage
$kurt$	Kurtosis of a random variable
L	Average number of users in system
L_A	Average number of users in system at user arrival instants
L_S	Average number of users being served
L_W	Average number of users waiting
M	Exponential distribution
M	Size of finite population
m	Number of servers
p_0	Probability that system is empty
pdf	Probability density function
p_k	Probability of k users in system
$q_1, q_2, q_3, \ldots$	Moments of time in system about origin
$r_{i,j}$	Transition (routing) probability
$s_{1(j)}, s_{2(j)}, \ldots$	Moments of service time about origin for jth priority class
$s_1, s_2, s_3, \ldots$	Moments of service time about origin
$skew$	Skewness of a random variable
T	Average time in system
$T_{(j)}$	Average time in system for jth priority class
T_A	Average inter-arrival time, operating time

T_S	Average service time
$T_{S(j)}$	Average service time for jth priority class
T_W	Average waiting time
$T_{W(j)}$	Average waiting time for jth priority class
u	Traffic intensity
$u_{(j)}$	Traffic intensity for jth priority class
var	Variance of a random variable
$w_1, w_2, w_3, \ldots$	Moments of waiting time about origin
z	Transform variable

Abstract

Convergence is undoubtedly revolutionizing communications technologies and businesses, thereby merging networks of different types and origins. The future of telecommunications depends on the success of combining old and new solutions to serve new customer needs. It is obvious that a new breed of technologies emerges to solve the fundamental differences between traditional telecommunications and Internet technologies, and to integrate present and future content services. The characteristics of these technologies are far beyond what has been currently deployed: they are based on open platforms, multi-interfaced, distributed in nature, componentized yet integrated, interoperate autonomously and spontaneously, and they are reliable and scalable performance-wise.

It is neither easy nor convenient to try to predict what the dominant designs of the era of converged and next generation networks (NGN) are and will be, but some trends can be observed as will be described in this book. Clearly, the most important trend of all is Intelligent Networks (IN) and IP integration and its emerging technologies. This integration takes place mainly in the service and switching layers, and thus accelerates the adoption of leading edge data communication technologies such as gigabit Ethernet, WDM, ATM, and VoIP. Another trend is the overall integration of access and transport network interfaces: there are various different appliances and thus access networks, but they are tightly integrating to each network operator's terabit core transport network. At the same time it is not only networks and services which are converging. Appliances from different vendors are increasingly multi-featured including voice and productivity applications, wireless data connectivity, and other sophisticated capabilities. It is clear that there will be a further consolidation of technologies in this area.

Consequently, this book addresses IN and IP analysis, design, and performance as well as their converging architecture for network services and signaling. It encompasses work on the integration of two different kind of networks (e.g., circuit-switched and packet-switched). This integration has started from simple interworking and interoperation of functions. And, it is targeting towards a fully converged network. The goal of our work is the design of the architecture for seamless interworking between different elements of the converged network (e.g., different access networks), which is necessary for the support of new hybrid services.

This book also examines current approaches towards the integration of signaling between packet-switched and circuit-switched networks. Until now convergence between the two networks has for the most part taken place in the transport and signaling layers. Signaling interworking architectures however cater for the specific class of voice services and do not provide a generic platform for service interworking with distributed services on the Internet. Through the adaptation of Parlay APIs a way is foreseen for homogenization of voice services over both Internet and PSTN while allowing the installed IN infrastructure to be used also for Internet telephony services and for service interworking between voice services and open distributed services in the Internet. The book describes a layered service architecture that can be used as a platform for service interworking.

What is the service - it is some abstractive framework, or the way in which we see its performance. What is the application - it is a way in which we implement abstraction into something real (e.g., software). Application is not program – it is an imaginary framework implemented (built) into software. So, how to create software of a service (application)? How to put abstraction into the framework? The first step is - divide abstraction onto blocks – what we want to have and what we want to see in this application, e.g., create a framework. The second step is - see what the real network/system can offer us, e.g., if it is able to support this application. And, the third step - is an analytical model. It shows and explains the operation of the application in the network, e.g. how the network supports the service(s).

To build proper and valuable models is the first, the main, and the most important task of every network services designer. The same was done in our research. According to that, our goal was not to model, analyze, and simulate processors' speed values from different vendors. For that purpose, we created an analytical model, which properly explains operation and interworking of various services in converged networks. And, how big the values of λ or μ will be - milliseconds, microseconds, or nanoseconds is not important if the balance equation $\lambda/\mu < 1$ (e.g., system stability) is satisfied.

Abstract (in German)

Die Konvergenz von Netzen und Diensten revolutioniert zweifellos die Kommunikationstechnologien und die Wirtschaft; sie vereint Netze verschiedener Typen und unterschiedlicher Ursprünge. Die Zukunft der Telekommunikation hängt davon ab, wie alte Lösungen mit neuen kombi-niert werden können, um Kundenbedürfnisse möglichst optimal zu decken. Es entsteht eine neue Art der Technologie, die grundlegende Unterschiede zwischen Telekommunikations- und Internet-Technologien auflöst, und in heutige und zukünftige Content-Dienste integriert. Diese Technologien basieren auf offenen Plattformen, sind mehrfach verbunden, verteilt in ihrer Natur, haben bereits integrierte Komponenten, interagieren autonom und spontan, und sind zuverlässig und skalierbar in ihrer Performance.

Eine Vorhersage über die vorherrschenden Designs konvergierender Netzwerke und solcher der nächsten Generation (NGN) ist äußerst schwierig zu treffen. Dennoch beschreibt diese Arbeit einige beobachtbare Tendenzen auf diesem Gebiet. Die IP-Integration und die Technologien, die aus den IP-Grundlagen entstanden sind, stellen zweifellos die wichtigste Tendenz dar. Diese Integration findet haupt-sächlich in den Dienst- und Serviceschichten statt und beschleunigt folglich die Einführung von führenden Edge-Datenübertragungs-technologien, wie zum Beispiel Gigabit Ethernet, WDM, ATM, und VoIP. Eine weitere Tendenz ist die Gesamtintegration der Zugriffs- und der Transport-schnittstellen. Es gibt verschiedene Endsysteme und somit auch Zugangsnetze, welche aber eng mit jedem Terabit-Kerntransportnetz eines Operators verbunden sind. Gleich-zeitig sind es nicht nur Netzwerke und Dienste, die konvergieren. End-systeme von verschiedenen Herstellern sind mit immer mit mehreren Funktionen ausgestattet einschließlich Sprach- und Produktivitätsanwendungen, Verbindung über Funk und anderen hochentwickelten Ressourcen. Es ist daher offensichtlich, dass es eine weitere Konsolidierung der Technologien auf diesem Gebiet geben wird.

Diese Arbeit behandelt IN, IP und konvergierte Netzdienste sowie die Analyse, den Entwurf und die Leistungsbewertung von Architekturen zur Signalisierung. Ferner umfasst diese Arbeit die Integration von zwei verschiedenen Netztypen (z.B.: Leitungsvermittlung und Paketvermitt-lung). Diese Integration beginnt mit einfachen Interworking- und Inter-operations-Funktionen und resultiert in einem vollkommen konvergierten Netz. Das Ziel dieser Forschung ist der Entwurf einer Architektur für ein lückenloses Interworking zwischen verschiedenen Elementen eines kon-vergierten Netzes (z.B. verschiedene Zugriffsnetze) zur Unterstützung von neuen hybriden Diensten.

Darüber hinaus untersucht diese Arbeit gängige Ansätze für die Signalisierungsintegration zwischen leistungsvermittelnden und paket-vermittelnden Netzen. Architekturen mit Signalisierungsinterworking sorgen jedoch für eine spezifische Klasse von Sprachdiensten und versorgen keine generischen Plattformen für das Interworking zwischen Diensten und verteilten Diensten im Internet. Durch die Anpassung der Parlay APIs hat man einen Weg zur Homogenisierung der Sprachdienste sowohl über das Internet als auch über ein PSTN-Netz gefunden. Dies wird durch die Nutzung von installierter IN-Infrastruktur sowohl für die Internet-Tele-phonie als auch für das Interworking von Sprachdiensten und offenen verteilten Diensten im Internet ermöglicht. Die Arbeit beschreibt ein Schichtenmodell für die Dienste, welches als Plattform für Dienst-Inter-working verwendet werden kann.

Was ist ein Dienst - er ist ein abstraktes Gerüst oder eine Art seine Performance zu sehen. Was ist eine Anwendung - es ist eine Art etwas Abstraktes in etwas Wirkliches (z.B. Software) verwandeln. Eine Anwendung ist kein Programm - sie ist ein imaginäres Gerüst, das in eine Software implementiert wird. Wie kann man daher eine Dienstsoftware (z.B. eine Anwendung) kreieren? Wie stellt man eine Abstraktion in ein Gerüst? Der erste Schritt ist, die Abstraktion in Blöcke zu teilen - was wollen wir haben und was wollen wir in dieser Anwendung sehen, z.B. Gerüst bauen. Der zweite Schritt ist, zu sehen, was uns ein reelles Netz/System anbieten kann, z.B. ob das System diese Anwendung unterstützen kann. Und der dritte Schritt ist ein analytisches Modell, welches die Funktionsweise der Anwendung im Netz erklärt.

Die wichtigste Aufgabe jedes Netzdienstentwicklers ist, passende und nützliche Modelle zu finden. Dem zufolge ist das Ziel, analytische Modelle zu kreieren, welche die Arbeitsweise und Zusammenarbeit von Diensten in konvergierten Netzen beschreiben.

1 Introduction

The telecommunication fixed and mobile, and data networks are going through the revolutional transformations in last decade. They are converging, resulting in new classes of services, e.g. hybrid services that span different network technologies. The development of these services is very important for the existing and forthcoming next generation networks because it allows using existing infrastructures, rather than requiring new mechanisms to be deployed [Jha06, Kry01, and Kry06c].

The dramatic development and growth of packet-based network technologies and services on one hand, and the high reliability and availability together with the simple service creation principles in intelligent networks on other hand, results in demand of new converged services classes that take advantages from both kind of networks. The IN and IP networks separately are far away from being an ideal environment for the new classes of services development and deployment. But, if joined together, they complement each other and create single powerful platform that extend each of these networks by using resources of other. And, it allows to multiply the amount of supported services (e.g., IN/PSTN services plus IP services, and hybrid ones) [Kry01, Kry06a, Mit07].

The Internet structure is flexible and open for new services creation and implementation. And, additionally, it is possible to use bandwidth more effectively, because of the IP packet-based nature. However, the security issues are still open.

In contrast, the intelligent network is much more complex and less flexible. But, it has a well developed call control model, and the security for its services support is not much of a concern at the moment, because of the absence of outside access, apart from through the signaling system No.7 (SS7) protocol which is secure enough. Once introducing other packet networks, particularly the Internet, and connecting to customer premises equipments, security issues come to mind [Kry03a, Wal00].

It is clear, that the future of telecommunication networks is packet-based. But, during the transitory phase networks have a converged nature. They use Internet-based protocols (SIP/SDP, PINT, SPIRITS), while keeping the IN services and call control features. The IN call control model is used as a model for call control using SS7, and the session initiation protocol (SIP) as well, whereby, the SIP protocol can be considered as an alternative to or a complementary action of the IN application protocol (INAP) functions in future.

The converged network service control layer supports both voice and data services in both network types. All intelligent functions are put at the service control layer. It provides all major requirements with respect to the network capabilities control, in order to make the service control functions independently from different kinds of networks.

Converged network supports services that fulfil the following requirements:

- Any combination of services: services have combine data, voice, and multimedia features and have to enable communication between persons, devices, and applications;
- Any configuration: phone-to-phone, phone-to-PC, IP terminal-to-phone, and so on;
- Any topology: one-to-one, one-to-many, many-to-many;
- Any user's choice of service or services set: the user profile is shared by any service, whereby additional preferences and constraints could be derived from the composition of profiles leading to a separation into groups and business relationships (e.g., user/ subscriber);
- Any terminal: the user has access to his/her services by means of any terminal (fixed phone, mobile phone, PC, PDA, etc. according to the terminal capabilities), in the same way and with the same subscribed features;
- Any application and interaction between applications: services can be activated by any communication applications (e.g., PINT/ SPIRITS services) and interact with applications, possibly deployed in third-party administrative domains (e.g., enterprises, added value service providers) [Kry07a, Min00].

Also, it is necessary to say that, integrated services, once envisioned in the circuit-switched world will come to fruition in the packet-switched world. It can be that the IP telephony of today will become the unplanned source of the ultimate integration of voice and data networks.

In the earliest history, packet telephony had poor voice quality and substantial delays. The first use of voice over IP was made with PCs at each end. The PCs introduced delay from a number of sources. In addition to the end point delays the best effort nature of the Internet introduced variable delays and even gaps in the transmission of voice that resulted in a perceptibly lower quality of service (QoS) than on the PSTN. Although there were some early applications of Internet telephony in 1996 and 1997, it was not until 1998 that services based on this new tech-nology became significant. The continued explosive growth of data capacity during this period took bandwidth used for data from one tenth that used for the PSTN to full parity. Not only the delay problem was resolved but also a whole new set of services came into being. The years 1999-2002 saw significant maturation of the IP telephony technology. The IP enabled phone came into its own in large part as a result of new services it offered that could not be matched by conventional phones [Rin99]. However, the ease of integration and scalability to deploy new services into the existing service structure is complex. The true value of a packet-based network will be found in the services that will be available through applications that exist on top of the IP network. And, finally, the benefits of circuit- and packet-switched networks convergence are many, but one of them is really obvious - next generation applications and services [Kry06b].

1.1 Categorization of Challenging Issues to be Investigated

We focus in this book onto the performance analysis, evaluation, and modeling of converged network services behavour. Also, the network parameters are investigated for the development of the architectures for new services delivery and support. The first step to that is addressing the existing challenges.

In converged network development some specific challenges exist:

- Convergence has the potential to reduce the cost of ownership, since the service provider can deploy all services on a single network rather than on a number of overlay networks. Nevertheless, this potential can be achieved only if the new network elements in the converged network are managed by an integrated management system.
- Although, support for voice communications using IP, VoIP, has become attractive mainly for its low-cost, flat-rate pricing of the public Internet, the technology has not been developed to the point where it can replace the services and quality provided by the public telephone network.
- In VoIP, the voice signal is digitized, compressed, and sliced into packets and sent with other packets across the packet-switched network. At the destination point, re-assembled packets arrive as normal sound voice call. Successful delivery of voice over packet networks presents a tremendous opportunity; however, implementting the products is not so straightforward with all the varieties standards, user requirements, interoperability, scalability issues, and still need research.
- The biggest challenge faced by voice over packet networks is that of providing the end users the quality of service that they get in a traditional telephony network. Unlike the PSTN, where a dedicated end-to-end connection is established for a call, packet-based net-works use statistical multiplexing of the network resources. Though sharing

resources amongst multiple users leads to a cost saving, it deteriorates the overall quality of service offered to a user. There are multiple parameters that determine the quality of service provi-ded by a network (e.g., delay, delay jitter, and packet loss).

- The various traffic types are associated with call holding times ranging from short (e.g., credit card verification) to long (e.g., Inter-net access calls). The distribution of holding times for the Internet calls in IN/PSTN has a long tail and a correspondingly large variance. For the traffic with these long holding times or a long-tailed distribution of holding times, the special observations may be required until estimates of parameters approach their true equili-brium (mean values).
- Multiple variants exist today for supporting voice over packet networks (e.g., voice over ATM, voice over IP and voice over Frame Relay). All these variants would have to co-exist with each other, along with the conventional PSTN. This raises multiple issues, but primary amongst these is the issue of interworking.
- Billing also becomes complex in this scenario, since the call traverses networks that differ in principle. In a PSTN network, a dedicated connection is established between users, and they are charged even at times when there is silence and no voice samples are generated. However, packet based networks charge a user on the amount of bandwidth used, and not on the total duration of the call.

1.2 Identification of Research Goals

The aspects of the real-time system design and the science of design are complex and important investigation themes. There is a lot of ongoing research in these topics separately but till now there is no simple methodology to bring them together. This book addresses some solutions for integrating them into a single platform.

The most important issue in every new modeling, analyzing, and evaluation study (e.g., modeling new systems, devices, networks) is to choose a proper theoretical (analytical) model. The correct choice gives precise calculations of main system characteristics (e.g., time in system, waiting time, service time, number of users in system and in the queue, and so forth).

The analytical model can show a first picture of the system performance. Also, it allows to see, if we were precise or not in our approximations and calculations. If we understand network/system performance properly then we can also evaluate where system/network strength is, and where weak points are.

For IN and IP telephony services, interworking is the ability to offer a broader service that results from their peer similarities comparison. The hybrid services provide connectivity between users of both networks as well as between users of the same network. Therefore, hybrid voice communications involve both IN/PSTN and IP services and/or both types of terminals. The goal of this research is building architectures for different communication scenarios in order to provide pure and hybrid services. The main objectives are as follows:

- Performance analysis of different evolutional processes in circuit-switched and packet-based communication networks to cope with the challenges that need to be addressed in the development phase;
- Performance analysis, comparison, and evaluation of existing protocols for signaling and control in pure IN, IP, and the converged network. And, based on this analysis, building and modeling of a signaling architecture for converged network services/applications support;
- The IP and IN areas represent a protocol concept and do not necessarily involve a real network. In this context, the development of interworking functions in a gateway is needed to perform all protocol conversions and data adaptations;
- Proposal of an architecture for pure and hybrid services support of basic communication scenarios;
- Analysis and selection of mathematical models, derived from teletraffic theory, for communication and services scenarios in the converged network architecture;
- Modeling of different service scenarios using methods of teletraffic theory;
- Performance evaluation of the proposed models and architectures and their protocols for implementing them into the converged network architecture;
- Based on analysis and modeling of service signaling message scenarios from one side, and the converged network and its capacity to support different services from other side, the evaluation of possibilities of existing and new services operating in converged networks.

1.3 The Book Structure

Chapter 1 is the introduction, where we bring in and explore chalenges, motivations, objectives, and our research contributions.

Chapter 2 presents an overview of the evolution and advancements in the service delivery with network intelligence enhancement. This process started from the simplest conceptual telephone networks, based almost on analog equipment. But, after the invention and development of new packet-based technologies and data communications, international bodies began to investigate alternative technologies for providing new services. These studies resulted in series of standards known now as the signaling system no.7 (SS7). They have paved the way for the intelligent network (IN) and a variety of its services. The next step in this evolution was the distribution of intelligence in telecommunication networks. And, this has allowed building open, distributed platforms for new services support. The convergence between fixed and mobile networks was the next logical step of intelligence evolution in networks and services. And lastly, converged IN, as a unified platform for voice, multimedia, and data; has been the final phase of network intelligence evolution. But, it has been an evolution process of pure telecommunication networks. Now, telecommunication networks are migrating towards Internet technology.

Besides, we perform in Chapter 2 the analysis of different signaling traffic flow processes in telecommunication networks, and apply queueing theory methods for the modeling of intelligent network services. The analysis is conducted using hierarchical decomposition techniques, allowing a detailed consideration of the signaling network protocol.

The 3rd Chapter concentrates on different integration aspects of IN and IP networks convergence. It gives a comprehensive survey of all standardization and research efforts in this direction. We analyze and evaluate different call models and the interworking between them for transparent service access from various underlying networks.

The most important features that describe every network are its services. Therefore, in Chapter 3 we discuss services that can be recognized as converged. We explain converged network services operating by signaling message flows for every service scenario. Further, we analyze and evaluate two protocols (PINT and SPIRITS) for the support of these services.

And as next important step, we analyze evaluate, and apply in this chapter, teletraffic theoretical techniques for interworking of converged network services interworking. We present and illustrate five basic scenarios for service interworking in converged networks in details. And, in Section 3.9.2, we investigate signaling protocols for different scenarios. Sections 3.9.3 to 3.9.7 deal with different queueing systems we apply to different service scenarios. We examine, assess, and motivate usage of every queueing system into every scenario implementation. Also, we provide numerical results of important system characteristics (e.g., time in system, waiting time, server busy period, number of users in system, number of users waiting in queue). And, we demonstrate the calculated results in numerous diagrams.

And, finally, Chapter 4 concludes our research work, by summarizing of goals and objectives we have reached.

2 Evolution and Advances in Service Delivery with Network Intelligence Enhancement

2.1 First Telephone Networks

As soon as the first installations demonstrated the feasibility of real-time communication between separated people, the demand for telephone service created the telephone exchange. And eventually, when automatic switches were perfected, the public switched telephone network (PSTN) was born [Car95, Schw88, Han99].

The simplest conceptual telephone network was consisting of two basic telephones with a simple transmission link between them. As this simple network expanded to include further subscribers, each subscriber acquired a direct link to every other, as well as a separate telephone for each link.

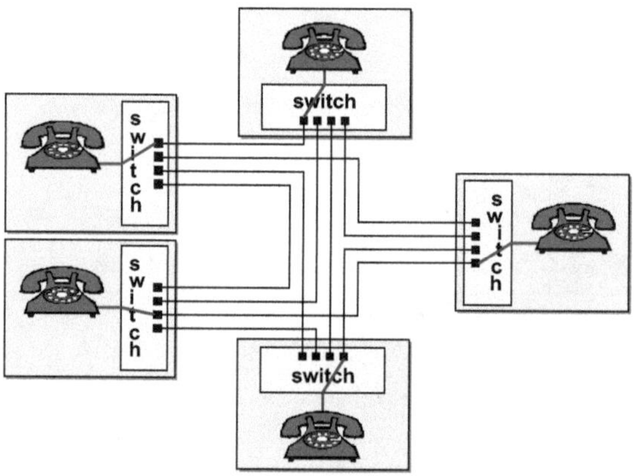

Figure 2.1 - A fully interconnected customer network.

The first refinement to this replaced the numerous telephones at a subscriber's premises with a single one. Some form of switching mecha-nism was then required to connect the telephone to the desired link (Fi-gure 2.1). This system was satisfactory if the subscribers remained few and close together; otherwise the number of links became prodigious and the cost of the fully-interconnected network prohibitive.

A network of practical proportions was created by concentrating all the switches in a central location – switching center, or exchange – and providing each subscriber with a single line to that exchange (Figure 2.2). Centralized switching offered equipment flexibility. Efficiency in providing and maintaining the transmission connections between subscribers and the exchange was also achieved. Instead of laying each connection, or circuit individually, a number which are all destined for the same area were put together into one cable.

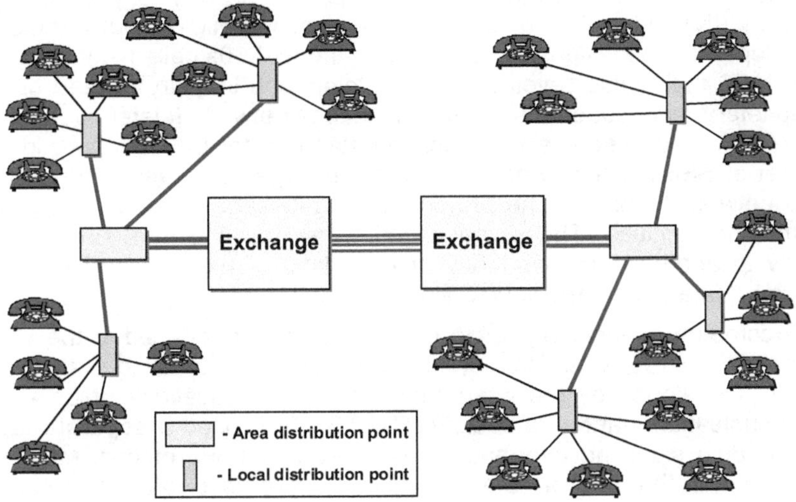

Figure 2.2 - First public telephone network.

This was provided, as an entity, between defined termination points, within the exchange at one end and a point of distribution to the local subscribers at the other. The circuit to a subscriber traditionally consisted of a pair of wires which formed an electrical circuit or loop between the exchange and the subscriber [Red95, Kühn94].

2.2 Common Channel Signaling Network

First telephone networks were the result of years of evolution. Based around analog equipment, the telephone network of the early telephone company was not well suited for services such as data and video. Many individual technology service providers began appearing during the 1960s, providing packet-switching networks and data communications services the telephone companies were just not equipped to provide.

The international telephone network was facing the same problems. In many countries, just getting simple international telephone service was a privilege. As international bodies began investigating alternative technologies for providing new telephone services, such as mobile, the need for an all-digital network became apparent. Thus, arose the beginnings of an all-digital network with intelligence. The international telecommunication union (ITU) commissioned then international telegraph and telephone consultative committee (CCITT) to study the possibility of an all-digital network. The result was a series of standards known now as signaling system no.7 (SS7). These standards have paved the way for the intelligent network (IN) and, with it, a variety of services. The ITU-TS (former CCITT) developed a digital signaling standard in the mid-60s called signaling system No.6 that would revolutionize the telephone industry. Based upon a proprietary, high-speed data communications network, it later evolved into SS7, which has become the signaling standard for the entire world. The secret of SS6 success was laid in the message structure of the protocol and network topology. The protocol used messages to request services from other entities. These messages were traveling from one network entity to another, independently of what kind of media they were (e.g., voice or data), in an envelope called a packet.

The common channel Signaling (CCS) was first introduced in the United States in the 1960s as common channel interoffice signaling system No.6 (SS6). Developed by the international telecommunications union – telecommunication standards society (ITU-TS), SS6 used a separate facility for sending signaling information to distant telephone offices. Messages were sent in the form of data packets and were used to request connections on voice trunks between two central offices. This became the first use of packet switching in the public switched telephone network (PSTN). SS7 was derived from SS6, but providing much more capability.

The procedure for tearing down a circuit is much faster in CCS than in conventional signaling, and is not as error prone. Even if voice circuits do get connected, with the speed of the signaling network, circuits can be disconnected and quickly connected again for a new call. While a call is in progress, information regarding the call can be sent through the SS7 network. CCS uses existing telephone company resources, so it does not require additional facilities to be installed. When signaling information is placed on existing digital transmission facilities, it uses a fraction of the circuits required for in-band signaling. One digital data link can carry the signaling information for thousands of trunks and maintain thousands of telephone calls. SS7 is the protocol and architecture used in this new network. There were many methods used for signaling but none of them could support network management functions or control information between switches and operations systems, except SS7. Because SS7 consists of a data network using data messages, SS7 can meet the demands of the evolving telephone network [Rus95, ITU-Q7, ITU-Q9, Duf94].

2.3 Signaling System No.7

Signaling system 7 (SS7) is the data communications protocol which provides the underlying network to support intelligent networking (IN). The deployment of SS7 and IN technology across the globe has been driven by the desire of telephone companies to offer new services, many of which will generate significant revenues.

SS7, also referred to as out-of-band signaling, is the common channel signaling protocol used for call handling within the telephone network and as the basis of IN. SS7 is the underlying data communications protocol used by telephone networks to control call set-up and call routing, and to provide services such as caller ID and CLASS features. SS7 offers telephone network management functions which are faster, more reliable, and more advanced than earlier technology by managing voice circuit functions on a separate, fully redundant data network.

SS7 was originally designed for exchanging call control information between the various network switches and databases of the PSTN. SS7 was increasingly used for new more complex purposes including enabling the deployment of new technologies such as ISDN (Integrated Services Digital Network).

2.3.1 SS7 Architecture

A telecommunications network consists of a number of switches and application processors interconnected by transmission circuits. The SS7 network exists within the telecommunications network and controls it. SS7 achieves this control by creating and transferring call processing, network management, and maintenance to the network's various components. An SS7 network consists of the following components (Figure 2.3):

- Signaling Transfer Point (STP): The STP is to the SS7 network what the switch is to the public switched telephone network. While a switch routes calls by making actual voice connections, the STP simply directs the digital traffic by selecting links on which to place the outgoing traffic. STPs are paired for redundancy with consideration being given that both members of the pair are not subject to the same hazards.
- Signaling Transfer Point (STP) - (Local): The STPs indicated here are at the lowest level of the SS7 network hierarchy. What makes them local STPs is the fact that the area which represents a network location (or node) providing and/or using network services is directly connected to these STPs. Just as a local telephone office is the direct connection point for the phone lines of telephone users, the lo-

cal STP pair provides the direct connection for users of the SS7 network.

- Signaling Transfer Point (STP) - (Regional): The STP pair indicated here is at a higher level of the SS7 network hierarchy. The drawing indicates this by showing no direct node connections, and also by placing this STP at a higher position on the page. The two pairs of STPs shown at the center of the network are, therefore, local pairs whose function is to provide network access to the services nodes. The higher level pair is the regional pair used to connect local STP pairs from different areas together.
- Signaling Point (SP): When a telecommunications service is to be connected to the SS7 network, it is given a signaling point code (SPC) identity much the same way a new telephone location is given a telephone number. A service with such a code is known as a signaling point. SP, however, is a broad generic term, which does not identify the type of service being offered. Other terms, such as service switching point (SSP), service control point (SCP), mobile switching center (MSC) and others, define the services offered in more narrow categories. For example, an SSP is a location offering voice circuit connections (in the telephone network) and SS7 connections for the exchange of circuit information and for call routing and maintenance requests. SCPs provide services such as database information, while MSCs control mobile networks and provide voice connections for subscribers.

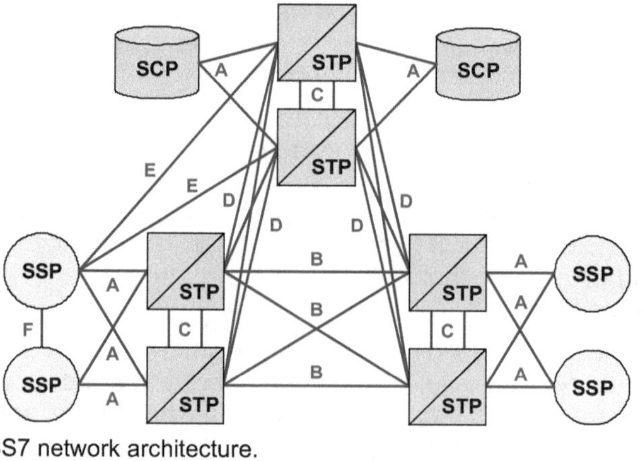

Figure 2.3 - SS7 network architecture.

What allows messages to travel around the network is the existence of connections between the nodes (e.g., signaling points) which are called data links. The SS7 network is unconcerned as to the type of transmission being employed except as it may impact the considerations of the physical layer of the protocol. Therefore, links are categorized by what they connect rather than how the data is transmitted. The names given to link types can be represented by the letters of the alphabet "A" through "F":

- Access Links: To provide services from the SS7 a node needs first to gain access. This is normally done through connections to a signaling transfer point (STP). STPs exist throughout the network on a hierarchical basis. That is, some exist for the prime purpose for providing access on a local basis to service providers (or service users). Other STPs may exist solely to expand the network by connecting local STPs. At a still higher level STPs can provide for international communication. The links that connect a node to a local STP pair provide access to the network, and are therefore called access links.
- Bridge Links: The more links available to an STP for connection into the network, the greater will be that STP's routing flexibility. To gain such flexibility an STP will link to a second STP at the same hierarchical level (e.g. Local to Local). The linking arrangement employed connects each of the STPs in one area with each of the STPs in the other area. To do so requires four links. Since these links form a bridge from one area to the other, they are referred to as bridge links.
- Cross Links: For the redundancy support, STPs are paired. In a redundant pair, it is generally assumed that both members of the pair perform exactly the same functions. Both members of a pair of STPs can be considered to be the same logical location. Since these links allow messages to cross over from either STP to its mate, they are called cross links.
- Diagonal Links: Even an STP linked to another STP at the same network level can gain additional routing strength by connection to an STP at a higher level (e.g. local to regional). The linking arrangement employed connects each of the STPs in one area with each of the STPs in the other area just as Bridge links do. In abstract terms, diagonal simply implies the connection of two levels of network hierarchy.
- Extended Links: While STPs are often connected to other pairs at the same level of network hierarchy, these links are commonly made to the closest pair on that level. Further routing flexibility can be achived by connecting to still another pair of STPs on that same level. To do so requires adding links to some more distant pair.

Such links would be made in the same four-linking arrangement as B links. Since these links form a connection to a more distant pair of STPs, they are considered to be extended further than other links and are, therefore, called extended links. One might also think of these links as extending the routing capabilities to the STPs.

- Fully Associated Links: From time to time, particularly in a proprietary network, users find it desirable to share data directly between nodes and to bypass intervening STPs. This is only done for nodes that are directly and completely associated such as those owned and operated by the same company. Since such linking occurs only between nodes with this complete association, the links are referred to as fully associated links.

2.3.2 SS7 Protocol Layers

SS7 standard was developed in a modular approach. This approach leads to the creation of what is referred to as a "layered" protocol. Protocol means nothing more than a fixed set of rules which determine how communication should be handled. It covers everything from what should occur to when and how it should occur. It also prescribes exactly what the message consists of when it is sent over the links. "Layered" means each module performs its function in sequence and then hands the message off to the next module (which is "above" for incoming messages and "below" for outgoing messages). Each of the functional program modules is termed as a "user part". The rules (protocol) dictate the sequence in which things must be done.

In Figure 2.4, the functional modules that deal with a message just about to be transmitted over the links (or one just received from the links) are shown at the bottom. Other modules are shown "stacked" above in the sequence in which their functions are performed. This picture is commonly called a "stack". And, it is typical SS7 stack.

- Message Transfer Part - Level 1: The message transfer part level 1 (MTP L1) is called the "physical layer". It deals with hardware and electrical configuration. It is necessary remember that a protocol is only a set of rules. Those rules extend to what occurs in the equipment to control the links. For example, one rule for MTP L1 is that a link must consist of two data channels operating in opposite directions at the same bit rate. In other words, the links must be bidirectional. The standard also refers to the need to disable certain attachments to the link that would interfere with full duplex operation and might challenge bit integrity. In other words, MTP level 1 is a user part that deals with physical issues at the level of links, interface cards, multiplexors etc. It does not, therefore, concern software providers except that they need to understand these requirements in order to interface the software module layers with the physical layer.

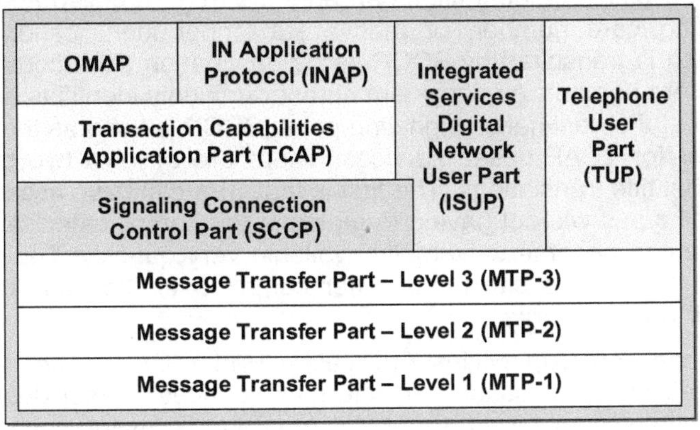

Figure 2.4 - SS7 protocol stack.

- Message Transfer Part - Level 2: This is a busy user part. It is the last to handle messages being transmitted and the first to handle messages being received. It monitors the links and reports on their status. It checks messages to ensure their integrity (both incoming and outgoing). It discards bad messages and requests copies of discarded messages. It acknowledges good messages, so the transmitting side can avoid of redundant copies. It places links in service, and restores the service links that have been taken out of service. It tests links before allowing their use. It provides sequence numbering for outgoing messages. And finally it reports much of the information it gathers to MTP Level 3.
- Message Transfer Part - Level 3: The MTP level 3 provides the functions and procedures related to message routing (or signaling message handling) and signaling network management. MTP L3 handles these functions assuming that signaling points are connected with signaling links. The message routing provides message discrimination and distribution. Signaling network management provides traffic, link and routing management, as well as, congestion (flow) control.
- Signaling Connection Control Part (SCCP): SCCP provides connectionless (class 0) and connection-oriented (class 1) network services and extended functions including specialized routing (GTT-global title translation) and subsystem management capabilities above MTP Level 3. Many of the benefits of the use of the SCCP lie in the

specialized routing functions. The addressing capabilities are what allow the locating of database information or the invoking of features at a switch. A global title is an address (e.g., a dialed 800 number, calling card number, or mobile subscriber identification number) which is translated by SCCP into a destination point code and subsystem number. A subsystem number uniquely identifies an application at the destination signaling point. SCCP is used as the transport layer for TCAP-based services. There are at least two benefits of global title translations. The first is that SPs can have access to data of all types without having to maintain too compli-cated tables. New data can become universally available very qui-ckly. The second is that companies can have better control over the data kept within their own networks.

- Transaction Capabilities Application Part (TCAP): The transaction capabilities application part offers its services to user designed applications, as well as, to OMAP (operations, maintenance and administration part) and to IS41-Cb (interim standard 41, revision C) and GSM MAP (global systems mobile). TCAP supports the exchange of non-circuit related data between applications across the SS7 network using the SCCP connectionless service. Queries and responses sent between SSPs and SCPs are carried in TCAP messages. TCAP is used largely by switching locations to obtain data from databases (e.g. an SSP querying into an 800 number database to get routing and personal identification numbers) or to invoke features at another switch (like automatic callback or auto-matic recall). In mobile networks (IS-41 and GSM), TCAP carries mobile application part (MAP) messages sent between mobile switches and databases to support user authentication, equipment identification, and roaming.
- Integrated Services Digital Network User Part (ISUP): The ISDN user part (ISUP) is used throughout the PSTN to provide the messaging necessary for the set up and tear-down of all circuits, both voice and digital. Wireless networks also make use of ISUP to establish the necessary switch connections into the PSTN. In the telephone network, ISUP messages follow the path of the voice cir-cuits. That is, ISUP messages are sent from one switch to the other where the next circuit connection is required. ISUP offers two types of services, known as basic and supplementary. Basic services consist of those services employed in the process of setting up and tearing down a call. Supplementary services consist of those servi-ces employed in passing all messages that may be necessary to maintain and/or modify the call. ISUP functionality can be further broken

down into 3 major procedural categories: signaling proce-dure control, circuit supervision control, and call processing control.

- Telephone User Part (TUP): TUP handles analog circuits only. In most regions of the world, ISUP is used instead of TUP for call management.
- Operations, Maintenance and Administration Part (OMAP): OMAP services are used to verify network routing databases and to diagnose link problems.
- IN Application Protocol (INAP): for IN services [Rus95, ITU-Q7, ITU-Q9, ITU-Q12, Duf94, Miz97].

2.4 SS7 Technologies for IN Services Performance Support

IN provides the backbone to support and define services to access all kinds of information. As the need for new features and services became more important to customers, the need to deliver those services and features in an economical way became equally important. The challenge facing telephone operators is being able to provide these features and services quickly and efficiently. IN makes it much easier. Services and features can be changed or deployed using simple procedures through a terminal, rather than through expensive programming changes made by certified technicians. All the customer needs is the facility (trunks) to utilize the new services.

The Intelligent Network relies on the SS7 network, which forms its backbone. SS7 provides the basic infrastructure needed for the service switching point (SSP), which provides the local access as well as an ISDN interface for the signaling transfer point (STP), which provides packet switching of message-based signaling protocols for use in IN and for the service control point (SCP), which provides access to the IN database. The SCP is connected to a service management system (SMS), which provides a human interface to the data base, as well as the capability to update the database when needed. The SMS uses a command-line interface and a man-to-machine language to build services and manage the network.

When a call is placed in the IN, a request for call-handling instructions is sent to the SCP using the transaction capabilities application part (TCAP) protocol. The database provides the instructions for handling the call based upon the customized service instructions the subscriber has programmed, and sends them to the end office switch. The call setup and release is handled using conventional SS7 protocols through standard interfaces between SSP and SCP. This standard interface is one of the most important items in the intelligent network. The reason for this is that it is the basic interface.

There are two basic parts of a communication link between two points on a network:

- Speech channel - on which calls are conveyed;
- Signaling channel - which handles control information for the cell, setting up a speech channel, releasing it, transferring numbers, and so on.

Signaling between exchanges can be carried out in two main ways, by using either:

- Channel-associated signaling (CAS), or
- Common channel signaling (CCS).

CAS was formerly the only method for signaling offered on networks. When digitization came into practice, exchanges were interconnected via pulse code modulation (PCM) transmission techniques. The first PCM systems were created according to the old CAS principle, but they eventually gave way to the CCS principle, which was more advantageous. CCS offers a more efficient use of signaling channels and the freedom to transfer call setup data and any other information via these channels. Moreover, as the exchanges are controlled by CPUs, it is very beneficial to establish a direct data communication channel between the CPUs in both exchanges that need to communicate, i.e., to introduce CCS links. Figure 2.5 illustrates the structure of SS7. The part that is of primary interest in relation to intelligent networks is the IN application protocol (INAP) within the component sub-layer of the transaction capabilities application part (TCAP) on the application layer, that is, layer 7 of the OSI model. Also, the telephone user part (TUP) is employed by existing net-works that use SS7 for normal speech traffic (normal calls). Networks can also use a modified TUP for communication between the SSP and the SCP.

2.4.1 IN Service Implementation Using SS7 with TCAP and INAP

Using TCAP/INAP offers numerous benefits. First, it provides a standardized protocol, which allows an open approach to different systems and differrrent vendors of intelligent network equipment. Second, and more important, it offers the capability of transferring call setup in the interface. Third, it provides more security in the communication from the protocols. Fourth, services that use devices to broadcast recorded announcements to subscribers during call setup, can use the signaling network to control the announcement device in the subscriber's own transit or local exchange by a remote SCP.

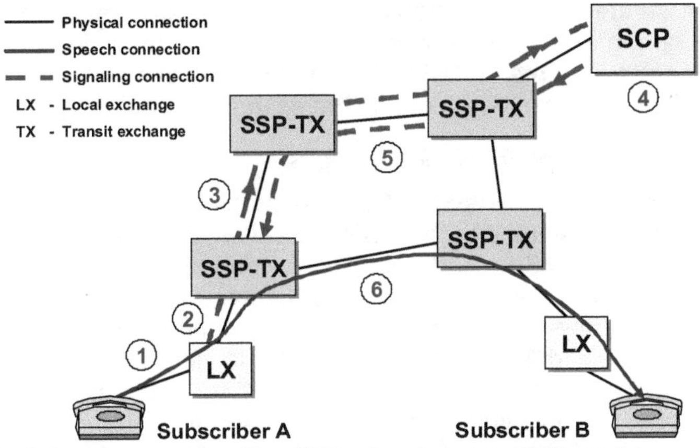

Figure 2.5 - A freephone setup with a SS7 network and an SCP solution.

Figure 2.5 shows a typical freephone service setup that uses the TCAP/INAP on a SS7 network. This solution allows signaling and speech to be separated; the TCAP/INAP works as a protocol conveyed by signaling channels, which, in turn, are regarded as a separate network (from the speech network), or a signaling network. The freephone service steps, illustrated in Figure 2.5, are:

- The local exchange hosting the calling party recognizes a freephone call from subscriber A by analyzing the called number.
- When it finds a freephone number, the local exchange routes the call (both speech and signaling) to an SSP, which is often the transit exchange.
- At the SSP, call setup is suspended while a setup to the SCP is made via the SS7 signaling network, using the messages conveyed by the TCAP/INAP.
- In the SCP, the freephone number is translated to the B number (i.e., the destination number for the freephone call).
- The B number is returned on the signaling network to the SSP that required the translation.
- When the SSP receives the B number back, normal call setup (signaling and speech) is resumed to connect to the destination local exchange and, finally, the B subscriber.

The TCAP/INAP is the way of signaling between SSPs and SCPs and, it is undoubtedly replacing all other methods. This is most important, of course, in implementations of mass calling services, like televoting, which can entail sudden bursts of traffic. But, the best solution is to have two protocols on the intelligent network, one that guarantees high security and performance of features (TCAP/INAP) and one that provides less security but offers fast response when that is the essential requirement.

2.4.2 IN Service Implementation Using SS7 and a Modified TUP

The normal procedure, when the SS7 has a TUP but not a TCAP/INAP, is to add some extra signals to the TUP exclusively for handling the communication between the SSP and the SCP. A freephone number setup with a TUP modified in this way follows exactly the sequence illustrated in Figure 2.5. The only difference is the method of signaling between the SSP and the SCP. A modified TUP solution uses the facility within SS7 that is offered by the quasi-associated or non-associated modes to create different paths for speech and signaling. Only the signaling channel is required, however, to transfer a number to the SCP for translation into another number, which must be returned to indicate where the freephone call must be routed.

Since the speech channel is not needed and the distance between the calling SSP and the called SCP could be very large, it is best simply to have no speech channel at all and to use only signaling. But this is not easily done in networks in which SS7 was mainly implemented to set up normal calls. Signaling requires a physical connection. There is a method, however, that can solve this.

The software in SSPs and in the SCP defines direct routes between all SSPs and the SCP. In practice, however, there are no physical connections, only hardware devices on both sides that are physically loop-connected back to the SSP/SCP. In other words, the SSPs and the SCP "believe" there are direct speech channels between them, but there are only loop-connected devices at both ends (see Figure 2.5). When a number translation is needed, the SSP at the calling party's end calls the SCP, just as if it were a normal telephone call, and a "speech channel" is reserved at both ends of the SSP/SCP connection (but no transmission link exists). As the speech channel (which does not physically exist) is reserved at both ends, signaling can be exchanged between SSP and SCP (dotted lines labeled 3, 4, and 5 in Figure 2.5). SS7 uses the quasi-associated or non-associate mode. When the number is returned to the SSP, the freephone call is set up, just like any normal call setup to B (the destination number).

The advantages of this solution are that it is easy to implement and takes less time to start up with (a simple) intelligent network, compared to the implementation of TCAP/INAP. It is usually not very performance-consuming in the CPUs and it provides short answering times from the SCP.

Disadvantages to the modified TUP solution, compared to the TCAP/INAP solution, are that the type of information that may be exchanged between the SSP and the SCP is very limited and that, after delivering a translated number, the SCP loses control over the rest of the service. This means, for example, that the SCP has no knowledge of the remaining part of the setup of the freephone number. It does not know if there was any answer, if the call failed because of congestion or another reason, and so on [Rus95, ITU-Q7, ITU-Q9, Duf94, Thö94, Lee97, Zna97].

2.5 Distribution of Intelligence

Distribution techniques within telecommunication systems establish the foreground for architectural implementation in heterogeneous environments for computational, contextual, and cooperative design sets. Intelligence in each of these settings provides the point and multipoint decision-making capabilities for operational evaluation and, quite likely, intelligent modification of the distribution techniques. Combined, the two methods afford today's and tomorrow's telecommunication networks the ability to operate in legacy, heterogeneous, and federated systems proactively.

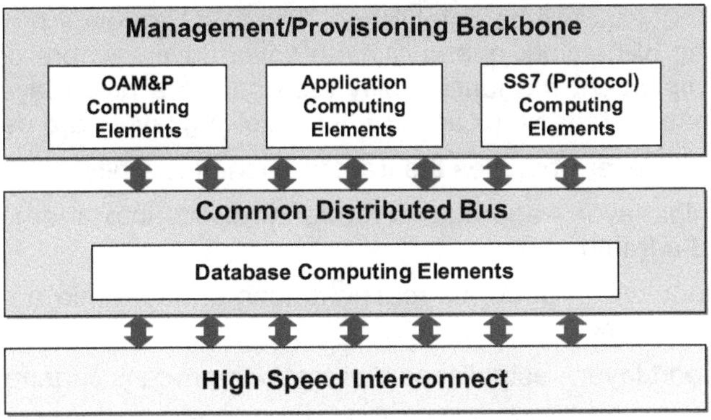

Figure 2.6 - Interconnected Intelligent Networking Responsibilities.

The distribution of intelligence in a telecommunications network begins as nothing more than segmentation of responsibility (Figure 2.6). The foundations of that segmentation are established according to the trend of moving telecommunications solutions toward more diverse computing platforms and away from monolithic settings. With movement and diversity comes the ability to integrate new solutions into the overall base system with greater speed and efficiency. Ultimately, the base system transforms to become part of a larger set of integrated components each with differring levels of responsibility and contribution to the intentions of that evolved solution.

Implementing the distribution technique requires several fundamental elements: a high-speed communication interface between participating computing platforms, a negotiated protocol between member services, and a delegation authority for assigning responsibilities to computing platforms based on the nature of their member services. These and many more decision-making activities continually occur in a capable system that dynamically acts and reacts to both the changing environment and changing needs of the networked solution.

Intelligence in the distributed environment finds its roots in the management of the solution. Cooperative behavior between member sets of the distributed environment lends data to the intelligent patterns. Most of all, the intelligent system grows. It exploits the diversity of the system topology to delegate responsibility to the outer reaches of the system informatively.

2.5.1 Fundamentals of the OSI Model

Distributed network intelligence is grounded in the OSI model. It is within this model that the necessary relationships between multiple hosts of the distributed network will be established. The OSI reference model defines a partitioning of network operability into seven layers where one or more protocols implement the functionality assigned to a given layer. Working from the bottom of the stack upward, the following layers are defined:

- physical layer - handles the transmission of raw bits;
- data link layer - aggregates a stream of bits into a larger data unit called a frame;
- network layer - performs routing among nodes within a packet-switched network;
- transport layer - establishes a process-to-process channel;
- session layer - creates and manages a name space used to link different transports considered part of a single application;
- presentation layer - provides a common format of data exchange between different types of peers;
- application layer - implements the functionality of the application.

Each layer provides services to the layer above in the protocol stack and uses the service from the layer below. For messaging, each layer adds its own header to the message being passed on by the layer above it on the sender side. At the receiver side, each layer takes off the header from the message and passes the unbundled message to the layer above it.

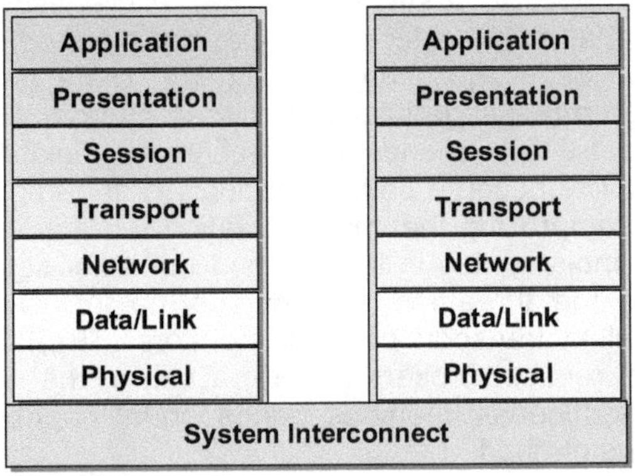

Figure 2.7 - The OSI Stack.

Figure 2.7 gives the graphical representation for the OSI stack. This model is replicated between disparate hosts. It is the replication that provides the framework for interoperability between compatible member hosts to execute in a distributed setting.

The OSI model provides a framework for the definition of the basic capabilities of an interconnection solution and strategy. The layers of the OSI model establish a simple method of rationalizing communication theories between disparate computing elements. For distributed proce-ssing techniques, the OSI stack extend the scope of basic inter-connection and opens up the discussion to include how the inter-connected elements may actually use and benefit from the connection mechanism at hand. Successful protocol agreements are achieved between distributed network elements through the use of the OSI model.

2.5.2 Distributed Processing Techniques

Distributed processing is more of a computer science concept than it is a telecommunications technique. Nevertheless, the extending of intelligence in the telecommunications environment encompasses the theories and principles of distributed processing. An information management and control activity, distributed processing involves the separation of work between computers that are linked through a common communications network. Methods of implementing distributed processing run between simple segmentation of workload and between member computer ele-ments to cooperative tasking of computer elements to achieve a singular goal. Common practices for implementing distributed processing take the form of client/server and distributed object architectures.

The client/server architecture arose during the 1980s as an alternative to centralized, mainframe computer architectures, although the client/server model can be applied to a single machine. In client/server methods, a client is identified as a requester of services and a server is identified as a provider of services. Negotiation between the client and the server is achieved through a message interface chosen by both the client and the server components. With client/server techniques, flexibility, inter-operability, and scalability become both byproducts of the implementation and requirements for improving the implementation. To this end, client/server techniques are implemented in two-tier and three-tier settings with variations applied to the three-tier methods for the inclusion of message-servers (also called transport protocol monitors (TPs)), application servers, and object request brokers (ORBs). Each of the transitional configurations for client/server methods brings about capabilities that build upon the previous method:

- Basic two-tier client/server implements simple request-reply actions in which the requester typically takes the form of an established graphical interface while the more powerful server actually implements the request and fashions a reply to the client/requester.
- Three-tier expands upon the limitations of two-tier architectures (typically sizing, processing overhead, and reliability) by implementing a logical middle tier that enacts message queuing. Maintained logically in both the applications of the client and the server, message queues are established to allow asynchronous operation on the client's part during the processing of the trans-action by the server.
- Transaction monitoring enhances the three-tier architecture by implementing higher orders of logic within the transactions themselves. This logic implements larger queuing methods that allow further abstraction of the asynchronous operations of the clients while at the same time performing redundancy operations that guarantee the safety of the in-flight transaction.

In ORBs, client/server architectures take on the evolving role of distributed objects. Distributed objects are the application of object technology to client/server systems. The architecture makes two distinctive presumetions: one, that the participating machinery in the architecture is capable of assuming and encapsulating the functions of an agreed set of common primitives known as common object services and, two, that the capabilities of object-oriented principles are available to the requesters of those services. The latter of these presumptions places the newly created services on equal footing with the basic primitives, which allows for larger and larger classes of services to be developed and integrated into the overall architecture's topology.

The application of distributed objects depends on the existence of a common set of services that is readily available to all participants in the system. These services are defined to be available through an interface known as an object request broker (ORB). The ORB allows other objects to make local or remote requests to other objects in the system. Primitive services available within the confines of the ORB are similarly available to local or remote requests at the same time as they cooperatively interact to provide the request/reply service. The object management group (OMG) has been instrumental in defining the components contained within an ORB. These components are the founding elements of the common object request broker architecture (CORBA).

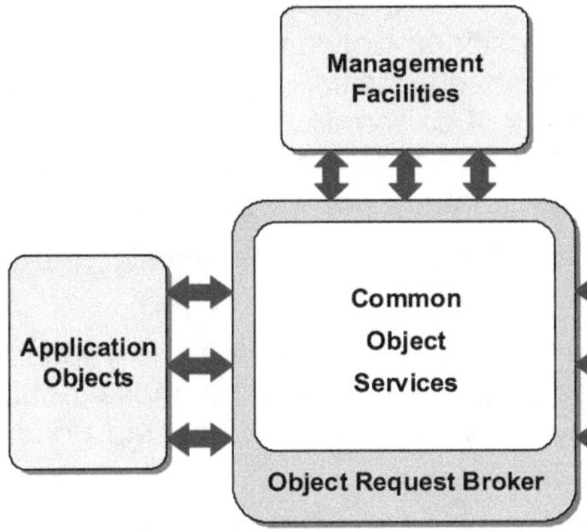

Figure 2.8 - CORBA ORB Architecture.

In this topology, the following elements may be found (Figure 2.8):

- an ORB that CORBA defines as the object bus, which allows objects in the overall system to perform request/reply actions to other objects in the system;
- the common object services, which provide system-level objects that allow the bus to interact with the system upon which it resides;
- management facilities, which refer to applications that are used by the application objects; the objects within this facility are generic to the overall system;
- application objects, which are the specific elements, that provide value-added work to the system.

One can find client/server techniques embedded within the distributed object topology. The important distinction made between client/server and distributed objects is established with the commonality that exists between member systems of a distributed object framework. While client/server relies upon either an agreed message interface or a mediation element such as a TP monitor, distributed objects rely on the existence of a common architecture from which an established mediation device maybe constructed. Distributed objects are an open architecture providing participation, development, and rapid deployment of solutions.

2.5.3 Aspects of a Distributed Network Intelligence

Distribution in the IN affords improvement at all levels of execution, operation, administration, maintenance, and provisioning. The main benefit found with distribution of intelligence is the ability to define systems that meet changing demands logically. A distributed system is one proactively designed for reactive behavior. In the IN, traffic loading is the principle reagent that influences the transitions between system states. Distributing the detection and reaction to state transitions between differing computing systems is an effective means of performing system modification while injecting the least amount of intrusive actions. In a distributed system, intelligent actions perform cures that are not worse than the disease.

At various layers of telecommunication systems, intelligent distribution occurs in several logic points:

- data/link (switching systems) implementations - dynamically allocate links or channels as nodes encapsulating those entities become available. Conversely, they de-allocate when the nodes are removed or altered.
- network implementations - perform dynamic routing and congestion algorithms based on behavior characteristics of participating elements. Such implementations route through or around nodes based on their current and predicted performance.
- transport implementations - mediate call flow between the objectives of the nodes to receive the calls and distinguish between node typing so that the appropriate call is enacted on the node that can best facilitate the objectives of the call.
- session implementations - correlate service provisioning to nodes capable of performing the service in question. Again, such implementations use the behavior patterns of the nodes in question combined with their ability to perform the service tasking to establish route paths to service nodes considered to be capable of performing the deployed service.

The general theme so far is to allocate to heterogeneous members of one's distributed IN those tasks considered relevant to the capabilities of those members. Configuration in this instance is an intelligent activity that dynamically changes as the nature of both the service requirements and system specifics change. This is intelligent behavior based on intelligent distribution. Perhaps the most commonly addressed distributed intelligent activity, however, runs a course through all of these activities. This is the action of performing network management.

Using the standard means of action/reaction to events within the system, network management works proactively to perform the traditional actions: configuration, event (fault), performance, provisioning, and security management. Each of these actions is triggered by behavior events in each of the participating systems. The network manager in this instance can either be an independent or participatory member of the system. As a result of the distributed nature of the system, the network manager beco-mes the vehicle for the overall coordination of state between the member elements to be able to define a single system state.

2.5.4 Diversity of Environments

A core competency of distributed systems and therefore distributed intelligence is the ability to relate tasking to the capabilities of the member nodes in the system. To this end, heterogeneous environments establish the best possible methods for applying intelligence toward a distributed system. In heterogeneous environments, tasking methods applied to the most appropriate container for the actions to be performed in the overall system are found. Distributed intelligence in heterogeneous models allows the system designer to accomplish the following tasks:

- retain use of legacy elements and systems;
- isolate usage of member computing elements;
- brand computing elements to perform best-fitting tasks;
- establish rules for functionality creep beyond designated computing elements;
- coordinate system behavior rules to overload escalation and abatement actions.

Using the client/server model for interaction, combined with the OSI stack as a requirement template for determining interactive behavior, one may begin to define the interoperability model between heterogeneous platforms, which accomplish the following:

- provide identification of the makeup of the member nodes or sets of nodes;
- establish a matrix of attributes to be applied to each member node;
- abstract the attributes to collections of nodes;
- identify the principal connectivity method between nodes;

- establish a common set of interactive primitives or messaging components between nodes;
- broadcast the attribute matrix to member nodes;
- establish an agreed manager of the system; voting ensues.

Now that the system is operable as a singularity, rules for reacting to changes in the system may be implemented either at the voted manager or within the limited member nodes. One has essentially established heterogeneous attributes as the common environment between the members of the system. Once the attributes are received and the method of system management (voting or otherwise) is established, each system can act independently to detect state changes and can then react corporately based on the rules for behavior dictated by the management scheme.

Change determines the behavior in distributed intelligent systems. One of the most predominant elements of change appears in system growth. For systems in which the preservation of assets is a key factor, growth most certainly brings about heterogeneous computing environments. Applying the principles of establishing best behavior for the appropriate elements in a heterogeneous setting, system designers are satisfied by the inclusion of legacy components in an evolving architecture. All that is needed is a redefinition and redeployment of the attribute and behavior matrix to the system. One must remove some of the functionality applied to the legacy node and distribute it to newly added members of the total system. The legacy environment is retained and continues to contribute to the activity of the overall solution. Several well-established solutions exist for enabling heterogeneous architectures. Those covered here are distributed computing environment (DCE), CORBA, and component object module (COM).

2.5.5 Methods for Intelligence Distributing

Building on the foundations of a distributed, possibly heterogeneous environment, it is discussed here the application of intelligent behavior to the components that comprise the system. The following two aspects of the system apply here:

- the system is distributed using object techniques.
- The system is performing a basic IN function.

Whether homogeneous or heterogeneous elements construct the system is inconsequential. The push is toward a distributed object environment working to provide intelligent behavior during tasking. Moving beyond the basic task of translation, many other elements of the system must operate simultaneously to provide availability, scalability, and manageability of the components and, therefore, the overall solution. Owing to these responsibilities, there are examined methods that involve message-based states, intelligent agents, rules processing, and state transition mana-gement.

- State transitioning - preservation of call-state during the flow of a transaction addresses one of the fundamental principles of OLTP (on-line transaction processing) techniques within intelligent networking. Call-state moves through several identities during a basic transaction's lifecycle. The net result is a much larger extension of basic request/reply models that are attributed to elementary client/server techniques. However, when examined at its basic components, each instance of the transaction's lifecycle can be interpreted as following the specifications of OLTP. Taken as a whole, the transactions themselves do not fall under the TP monitor's initiatives (i.e., there are no back-out activities). Instead, the trans-action itself carries with it the notion of call-state so that the steps performed during transaction processing can take that state value into account. Moreover, persistent operations administration, maintenance, and provisioning (OAM&P) activity that takes place during transaction processing can enter into or be enacted by the state transitions. These, too, involve actions determined by the current state that invoked the OAM&P functions. Thus, looking at the transaction lifecycle as a series of iterative operations uncovers a series of operations connected to each other both with data and with state value. These operations are then correlated to actions dictating the behavior of the next iteration in the lifecycle and the behavior of the overriding OAM&P activity. Two common methods are used to manage the state transition: callback methods and message-based methods. For true distribution techniques, the message-based methods prove to be the more appropriate vehicle.
- Callback methods - in an application program or a transaction- processing environment, a callback is a user routine or method that is registered with a state engine. When registered, the name or address of the callback is provided along with a triggering mechanism recognized by the state engine. The state engine retains the function/method address and invokes that address upon enco-untering the registered event that corresponds to the triggering me-chanism. Callback message-processing methods offer a distinct ad-vantage in that they centralize all message-processing code while, at the same time, they work with a uniform state value. Multiple call-backs can be registered with a single event either at the registration point (with the state engine) or as a second- and third-tier operation. For example, the primary callback could instigate a series of functions or callbacks. Rather than performing event-trigger registra-tion with the

state engine, different functions could register with the master callback. A callback model implies a multithreaded model.

- Message-based methods – message-based methods for transaction processing give the application program greater control of the timing of actions to be performed during the call's lifecycle. The method requires a generic set of actions used to check the content of the message to the point that interests the particular function receiving the message. The parsing methods increase in depth of discovery as the transaction moves through the system and builds upon itself. This is required as the transaction itself carries its current state as it moves along. Additionally, OAM&P activities may be generically invoked along the transaction's path to determine (again, based on the parsed combination of state and value) if actions should be taken outside the normal call path. One easy example is the notion of discarding aged transactions. A generic routine, which examines the current time and compares it to the originating time of the transaction, can be provided. At each point along the transaction's iterative movement through the system, the routine is invoked. If the delta between the two timestamps ever moves out of acceptable range, the transaction processing is halted. This does not require registration or notification because the triggering mechanism itself is a logic test-not an event within the state engine. Building upon this principle, generic logic gates can be developed that add intrinsic management and reporting values to the entire transaction path.
- Intelligent agents - building upon the discoveries in the computer industry surrounding artificial intelligence, agent technology has emerged as a significant contributor to providing user-level understanding of flexible and dynamic computing environment conditions. As previous sections have clarified, an intelligent networking environment is by far a dynamic environment, and intelligent agents in the middle of the transitioning activity provide a brief method of proactively and reactively changing the behavior of the overall system. Loosely defined, an agent is an entity (typically software) that is able to perform actions in a dynamic environment and that ultimately serves the directions of its creator. An agent is classified by its basic behavior, instantiated by the creator. Several classifications exist.
- Mobility - either the agent or its underlying behavior is static to a particular element of a network, or it is declared as mobile, indica-ting that it is able to transverse individual computing elements of the network. The latter implies a series of computing elements able to service the coding specifications of the mobile agent. Clearly, Java

comes to mind in the form of acceptable coding environments, as it involves platform-independent behavior.

- Intent - fundamentally, an agent's behavior is either deliberate or reactive. A deliberate agent tends to possess a set of actions to be performed in the form of an internal set of rules and or goals to be achieved. Through collaboration with other agents in the system, a deliberate agent acts to modify system behavior. Reactive agents, on the other hand, contain an orderly set of actions to be taken based on the changing nature of the system.
- Personality - performs actions on a system either through collaboration with other agents in the system or as an independent autonomous event. A collaborative agent determines its behavior and implements its actions through negotiation and conversation with other agents in the system. An autonomous agent typically gathers its information and affects the system directly.
- Roles - one of the easiest methods of classification for agents is the role that the agent plays in the system. Information agents, as an example, are associated with fundamental information-gathering activity. Reactive agents extend the role of an information agent by performing an action based on the content of the gathered information. When placed within the boundaries of an IN, agents serve an immediate purpose in the correlation of activities traditionally found within OAM&P. The trio of events, statistics, and overload offers the best example. Events, statistics, and overload interact with one another and ultimately influence each other's behavior. Events are triggered, which subsequently poll statistics, which eventually escalate or decrease an overload condition, which then triggers an event. Tying the three together and then attempting to perform corrective behavior in the form of process or application management becomes a challenging task from a static-system-description point of view. Interoperating agents between the components with rules of engagement based on the reactive and collaborative behavior softens the impact of the changing system. Furthermore, adding discovery-oriented agents to negotiate with the trio allows the end user the ability to perform off-line intelligence gathering about the dynamics of the state change. Through proper information passing, an explanation for various events can be determined.
- Rules-based processing - rules processing forms the user's mandates for specific conditions and actions to be detected and performed by agents during system environment transition. Fundamental rules take on the form of if-then-else constructs. For example, if a link goes out of service, issue an alarm. Or, if the link is out

of service, route to another link. Implementation of a rules-based method within an agent setting implies continuous checking for the enabling of the rule followed by actions. When compared to prior discussions on callback methods for state transitioning, the rule itself can be registered as a state within the overall engine. So the difference here is the addition of the state change registration thro-ugh the when condition. Architecturally, the result is an abstraction of a state engine-callback environment held above the fundamental message processing. The differences occur in that the agents manage the state engine through callbacks enabled and executed outside and independent of the call path.

- Interaction - Rules allow the agents to tie the models together. Through rules-based processing, agent and non-agent entities in the call flow are able to perform scripted actions based on registered occurrences. In many cases, the registration of events to detect is also determined through rules logic. An intelligent agent in the call-processing path is secure to a wide variety of information regarding the activity of the system and, using this information, is able to interact with other agents to effect change. It is the act of change that is determined by rules scripting. State table management as an overriding factor in the call path(s) of the system allows for abstract implementation of dynamic rules processing. Agents may implement a callback method for notification of the when clause of the rules script. The state table becomes the mediator of scripted agent behavior, while the call path remains the stimulant of agent activity [IEC-DNI, Hac98, Tai99, Ach02, Zui96, Por97].

2.6 Platforms Supporting Distribution and Openness

The telecom service providers tend to constantly offer more advanced services. Service uniqueness is a means of success in a competitive telecom market. The intention of intelligent networks is to provide means to meet these challenges by offering a platform for service development and execution. This goal has to some extent been met, and ongoing standardization indicates that IN tries to keep same speed with the growing demands. Still, between one and three years are probably required for successful introduction of a quite new tailored service. The main reason for this is that the SCP manufacturers have failed in providing platforms supporting standard tools, multi-protocols, and standard databases.

A serious problem caused by the lack of tools, and thus specialized skills required to developing IN services, is poor programmer availability. An open platform is the only means to decreasing cost and time-to-market for new IN services.

2.6.1 *IN Implementation – Hypothetical Travel Agency Service*

This section describes a service for handling customers calling a travel agency (TA). It is assumed that the service was originally developed and available for PSTN only. It is shown how the service may be extended to include interfaces to additional access networks like mobile networks and the Internet. It is also shown how to outsource the execution and development of the service to an external service provider interfacing to the experimental platform using a standardized interface like.

The TA has one toll free phone number, and a number of offices handling calls to this number. Each office has a number of travel agents. The caller is first exposed to a menu listing a number of options, e.g. domestic travels, travels abroad, holiday travels, etc. Each option is associated with a digit to be entered by the caller to indicate the topic of the request.

Each option in the menu is associated with a dedicated queue and a list of offices able to handle calls concerning this topic. Thus, an office is homogeneous in the sense that all agents in the office can handle the same type of requests. After selecting the most suitable option, the call is first attempted transferred to the offices in the list associated with the queue. If all agents in all relevant offices are busy, the call is inserted into the queue. Regularly, the first call in the queue is attempted transferred to the offices.

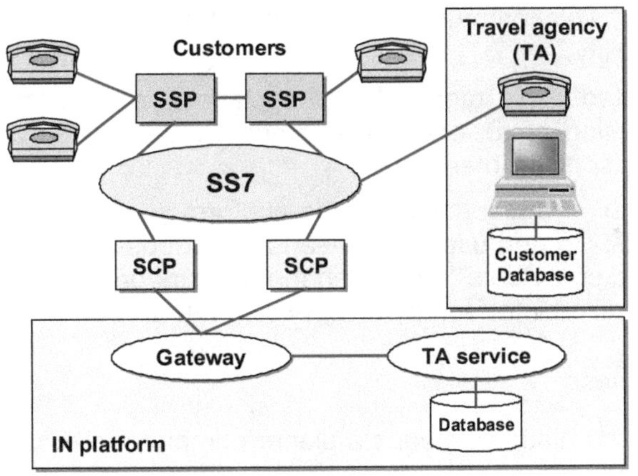

Figure 2.9 - Hypothetical travel agency service.

When the caller is eventually connected to an agent, the two parties agree on a product and a price. The caller may provide his credit card number to pay in advance, or he may choose to be invoiced. Information about regular customers may be registered in advance and stored as customer profiles at the TA in a database.

A possible overall architecture of the travel agency service is depicted in Figure 2.9. There are basically three independent parts in this architecture:

- The customers, PSTN and the SCP.
- The IN Platform with the gateway to PSTN, the service for queue handling and a database to store the queue.
- The travel agency, possibly with its own customer database.

The service described here is very basic, akin to what many companies have. In the next sections, it is explained how the service may be extended in various ways, with respect to functionality as well as how to outsource the service to an external service provider, and finally how to open up the platform for additional access networks.

2.6.2 IN Service Platform Requirements

An experimental platform for IN services should be:

- Open - in this context, this is related to the concept of using standard hardware and software as platform building blocks.
- Reliable - this is a very broad term and includes both availability and security. Telecommunication services traditionally have a high level of reliability.
- Distributed - this means that the platform should facilitate the decomposition of a service into autonomous entities communicating through specific interfaces.
- Accessible - this means that the platform should be available in two respects: To end users, by means of various access networks, and to service providers (other than the network provider), by means of a standardized interface to the service platform.

2.6.2.1 Openness

An open service platform means a platform implemented using widely applied, standard, commercially available hardware and software components. There are a number of reasons why it is considered this a requirement to the platform:

- Applying widespread technology means that the number of alternative solutions to choose from is high, and that a suitable product in terms of cost and quality can be purchased.

- Hiring qualified and skilled people to develop and maintain the service platform and the services is easier since more people can be assumed to have experience in the technologies applied.
- Interfacing the platform to other entities such as additional access net-works and services provided by external providers is easier.
- Keeping up to date with the technological evolution is easier because new technology is likely to be based on the most promising and widespread existing technology.

2.6.2.2 Reliability

The platform should be fault tolerant, i.e. be able to recover from failures without performing incorrect actions. Recoverability means that failed components are able to restart themselves and rejoin the system after the cause of failure has been repaired. There are two related aspects of reliability:

- Availability - the platform should provide high to continuous availability. This means that the platform should be able to resume providing services during recovery from failures. By minimizing recovery time, the platform is capable of providing virtually uninterrupted service to its users.
- Security - security protects an information system from unauthorized attempts to access information or interfere with its operation. Since the platform should be accessible by both subscribers and external service providers, a multitude of security issues should be investtigated:
    - identification and authentication of both humans and objects;
    - authorization and access control;
    - confidentiality;
    - security auditing;
    - security of communication;
    - accountability and non-repudiation.

Accountability and non-repudiation are related terms. Users are accounttable for their security-relevant actions. Non-repudiation provides serious evidence of actions such as proof of origin of data to the recipient, or proof of receipt of data to the sender to protect against subsequent attempts to falsely deny the receiving or sending of the data.

2.6.2.3 Distribution

Distributed object technology is well-suited for creating flexible systems because the data and business logic are encapsulated within objects, allowing them to be located anywhere within the distributed system. This is beneficial in terms of providing a logical decomposition of the system into independent units with clear interfaces, but also because the logical separation of entities easily can be extended to a physical distribution as well, thereby facilitating replication and increased reliability.

Another benefit of distribution is that scalability is supported since the system easily can be augmented with a number of additional components of a certain type. This of course assumes that there is a mechanism for distributing system load among numerous components of the same type.

Distributed objects separate their interfaces from the implementation. They are able to describe their interfaces, events and properties "on the fly". The implementation language of server objects is transparent to clients, which enables the use of distributed objects as wrappers for existing applications regardless of implementation language. Existing IN services should therefore smoothly migrate to the new platform.

The basic idea of the distributed object middleware architecture is the object bus, which e.g. in CORBA is the object request broker (ORB) that lets objects interoperate across address spaces, languages, operating systems and networks. The bus provides mechanisms that let objects exchange meta data and discover each other. At the next level, the infrastructure augments the bus with system-level services including licensing, security, version control, persistence and transactions. Thus, many important low-level services, previously implemented by means of proprietary mechanisms, are now available as parts of off-the-shelf comer-cial middleware systems.

2.6.2.4 Accessibility

The experimental platform should provide a standard interface for access by external service providers (ESP). The Parlay group (AT&T, BT, Cegetel, Cisco, Ericsson, IBM, Lucent, Microsoft, Nortel Networks, Siemens and Ulticom, Inc) focus on the production of an API specification that enables enterprises outside of the network domain to access network information and control a range of network capabilities. The published Parlay API has quickly gained popularity for this purpose, and is briefly described below. Figure 2.10 shows where the Parlay API fits into the architecture. The figure and the description below are extracted from.

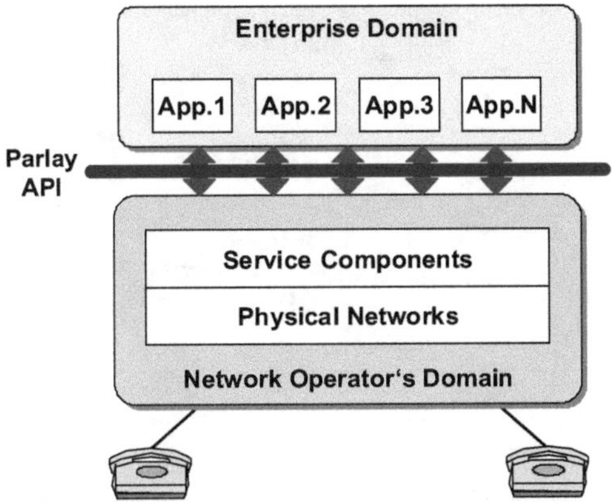

Figure 2.10 - Parlay API.

The Parlay API consists of two categories of interfaces:

- service interfaces - offer applications access to a range of network capabilities and information. Examples of the supporting functionnality provided via the framework interfaces are: authentication, discovery, event notification, integrity management, and finally operation, administration and maintenance.
- framework interfaces - provide the supporting capabilities necessary for the Service Interfaces to be secure, resilient, located and managed. Examples of service interfaces are: generic call control service, INAP call control service, generic messaging service, generic user inter-action service, and call user interaction service (voice prompt to user, DTMF input from user).

2.6.3 Open Service Platform Architecture

Figure 2.11 shows the open service platform architecture. It interconnects through several interfaces to different kinds of access networks (currently SSC, X.25, TCIP/IP and CORBA). The platform is both scalable and fault tolerant.

All items within the quadrangle are part of the platform. Circles denote functional CORBA-aware units. Stacked symbols indicate replication, and replicas always run on separate physical hosts for fault tolerance and load sharing.

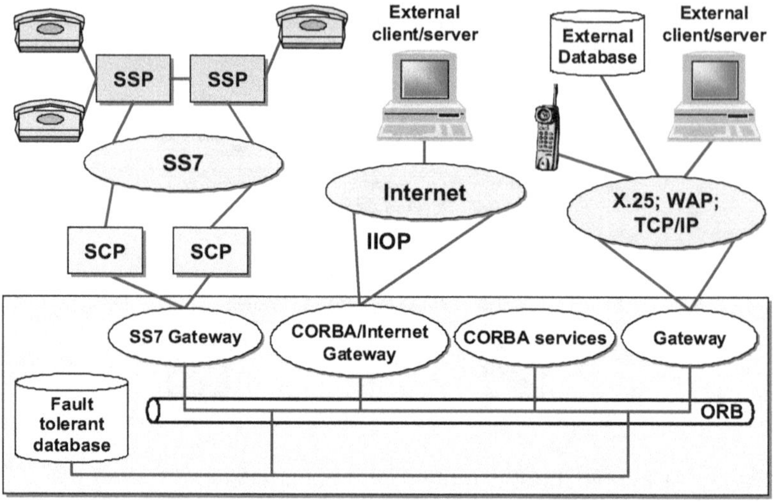

Figure 2.11 - The IN service platform.

In Figure 2.11, gateways (GW) are generic (i.e. service independent) protocol converters, converting requests and responses to/from CORBA. Some of the gateways also have routing and firewall functionality, e.g. to overcome the Java "sandbox" limitations. The database is a standard SQ1 Server. It is used by the services to store state between requests belonging to the same dialog. A fault tolerant name service uses "hot" replication based on group communication. Based on this platform there are implemented a variety of services, including:

- Traditional IN services (miscellaneous forms of call routing) with web interfaces for customization.
- A service creation environment consisting of a generic set of building blocks for creating call center solutions.
- An electronic commerce service, with external interfaces for customers, merchants and banks.

The functional units of the architecture of the platform are:

- SCP - implements a generic script which simply passes INAP requests to the SS7 gateway. This design allows new service deployment without SCP MIB programming.
- SS7 gateway - CORBA-aware gateway converts SS7 INAP requests to CORBA, analyses the message and hands it over to an instance of the appropriate replicated service objects. Upon reception of an answer from the servicing object, the reverse action is per-

formed. Missing response from the requested object causes retransmission to another instance of the desired process, residing on another application server.

- Middleware - CORBA represents the platform middleware, i.e. the runtime environment or object transaction manager (OTM) for the service implementation objects or components.
- Application server - service object implements service specific logic. This facilitates the introduction of new services by deploying new service objects on the application servers.
- Database server - internal database hosts data that needs to be stored during service execution, e.g. data about the state of service progress. It can also be used to host data for a service that runs solely on the platform, and therefore has no external database.
- Gateways - are developed on demand. Of particular interest are gateways to:
    - the Internet;
    - mobile terminals through WAP/WML;
    - external service providers.
- Fault tolerant naming service - connections between clients and servers within the platform are permanent in order to avoid object binding for each service request. In the case of failure, some rebin-ding is required. The fault tolerant naming service enables rebinding of all connections upon failure detection instead of rebinding within a service request. This service is implemented by means of a re-plicated supervisor process. The supervisors implement the replica control protocol for group communication.
- Servant objects register to the replicated name service by providing kind (type) to identify redundant objects.
- Clients request object kind in order to retrieve an IOR to any object of requested type, followed by an optional register interest message to the supervisor. Clients commit to reporting bad references to the name service, which in turn notifies all interested clients about the bad reference and provides a new IOR.

Since it is already identified several requirements to the platform for telecommunication services, it is of course relevant to discuss the proposed experimental platform according to these requirements.

- Openness - the platform is based on CORBA, which is a comercially available system for communication in a distributed environment. A standard language is used for defining object interfaces. CORBA is language independent.

- Reliability - CORBA provides mechanisms to handle application errors through distributed exception handling. Additional services to restart failing components may be implemented based on CORBA.
- Availability - fault tolerant naming service ensures that object references are valid, even in the case that the server fails. All servers are stateless. If a servant object needs to store state between invocations, it will use the (fault tolerant) database.
- Security - CORBA provides basic mechanisms for secure communication by means of Secure Sockets Layer (SSL).
- Distribution - CORBA provides the necessary means to define clear interfaces in a programming language independent manner. Thus, components of various implementation languages may communicate. Components on different ORBs may communicate through the Internet Inter-Orb Protocol (IIOP). Thus, physical distribution is well supported.
- Accessibility - Gateways may be implemented to provide interfaces to additional access networks as well as external service providers. The versatility of CORBA makes it possible to implement the gateways in a number of languages.

The last item concerning interfaces to the service platform is one of the most important aspects of a modem service platform.

2.6.4 External Interfaces

A possible extension to the service is to add capabilities for online payment processing. Thus, a customer can choose to register himself with the travel agency by providing information about credit card numbers, bank accounts etc. This therefore becomes an extension to the previously existing customer profile.

After the conversation with the agent is finished, the call can be transferred back to the TA service, and a payment dialogue can be initiated. This assumes that there is a connection between the travel agency computer and the service, to let the TA service be notified about the payment details. An architecture supporting this functionality is depicted in Figure 2.12. Here, the TA service provides more functionality than the service introduced in Figure 2.11 since it also has to support processing of payment transactions and communication to external banking services etc.

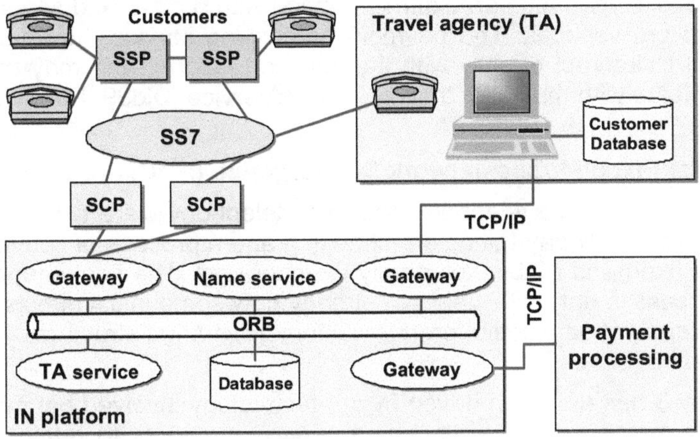

Figure 2.12 - Extended functionality of the TA service.

The overall architecture for a service platform providing access to external service providers (ESP) is shown in Figure 2.13. The interface is built according to Parlay on top of CORBA, thus requiring a CORBA-enabled external platform.

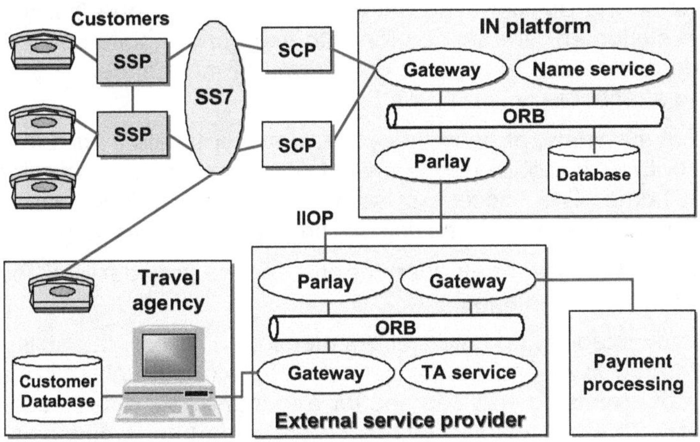

Figure 2.13 - Parlay/IIOP interface to external service providers.

The Parlay interface allows service requests to be passed to the external service provider. Similarly, the ESP is able and allowed to perform PSTN switching. The entire TA service may in this way be developed and maintained by the ESP. In the architecture above, it is assumed that the payment processing is also an external service with a clear and secure interface to client services. The payment processing service, however, does not need to interact directly with the service platform, and may therefore communicate with the caller through the TA service [Did99, Parlay].

2.7 IN for Fixed/Mobile network Convergence (FMC)

Until recently, mobile telephony and fixed telephony were two clearly separated worlds. Today things are changing and a process of convergence between fixed and mobile telephony is under way. The final shape of the FMC process is not quite clear but already now some initial effects can be seen in terms of services offered to customers, market structures and network infrastructures.

So far FMC has not been driven by the technology involved but mainly by the need for network operators and service providers to retain existing customers and attract new ones in an increasingly competitive market. For network operators and service providers the ability to provide integrated fixed and mobile services may be a winning card to obtain a new market share and to be able to distinguish themselves from competitors.

From a customer's point of view FMC means an integrated service package, whereby the customer is offered both fixed and mobile services, using one terminal and possibly one number, and receives one bill. It also implies one customer care service for both fixed and mobile services. FMC meets the customer's expectations in terms of independence from a terminal and a technology.

In terms of terminals, already today there are terminals available that can be used both as a mobile phone away from home and as a cordless fixed phone at home. The success of some commercial offers based on this kind of terminal seems a promising starting point for development of FMC.

From the point of view of an operator or a service provider FMC potentially has a double advantage.

The first advantage is commercial. As mentioned before, the ability to provide integrated fixed and mobile services offers opportunities both for the provision of promising services and for a recognizable differentiation from competitors. In other words, FMC can be regarded as a mechanism to increase revenues per subscriber and to gain new subscribers. In particular, fixed and mobile operators can integrate each other's market shares. For fixed operators the convergence may slow down the growth of the mobile market and give vitality back to the fixed market. For mobile operators FMC is a way to compete in the new sectors that are dominated by fixed operators such as data and value-added services.

The second advantage is operational. FMC may allow network operators to save money in terms of infrastructure and operational costs. The idea is to run one single network instead of two separate networks, one for fixed services and one for mobile services.

2.7.1 General Regulatory Aspects

FMC is taking shape, promising to be initially a good commercial opportunity and in due course a sound technical solution. At the regulatory level not much has been done so far and regulatory convergence is still far from being achieved.

Currently in most European countries mobile operators and service providers are under a regulatory regime which is much more open and free than the regime for fixed operators. The obligation to provide number portability is a clear example of this. Number portability in fixed networks, in line with the EU directives, is available in most EU countries. In the mobile environment number portability is a relatively new concept and its availability on a large european scale is years away.

Another example of the imbalance in regulation between fixed and mobile is the different approach, in some european countries, in terms of obligetion to provide universal services, indirect access and interconnection rates based on cost orientation.

This difference between the fixed and mobile regulatory regimes can be attributed to two historical factors:

- the initial need to facilitate the take off of mobile services, creating the right conditions to attract investments to roll out mobile networks;
- the fact that within the mobile sector there is not a situation of dominant operators able to exploit their position to distort competition.

Today the situation is quite different. The mobile market is well developed and is directly competing with the fixed market in terms of services offered, revenues and customers. According to some forecasts, before the year 2010 in Europe, the number of mobile customers will exceed the number of fixed customers.

In addition, falling prices for mobile calls are creating a situation of call substitution where instead of making a call using available fixed telephones customers tend to use their mobile phones.

In order to prevent forms of distortion of competition between the mobile and fixed sector it is crucial to promote a common regulatory framework. This common regulatory framework cannot be just a transposition of the “fixed” regulatory regime into the mobile sector. It is expected that the common framework will be based on light regulation able to guarantee safeguarding of customers' rights and fair competition between the parties concerned without abuse of dominant positions. Regulatory intervention must be envisaged when the market conditions do not meet criteria for fair competition.

2.7.2 E.164 Numbering Regulatory Aspects

To identify regulatory E.164 numbering aspects, different elements of combined fixed-mobile service offers should be distinguished:

- Integrated customer care and integrated billing - customers have a single contact point for customer care and receive a single bill for both their mobile and fixed terminal usage.
- Single handsets - customers have a single handset for fixed and mobile services. As long as the fixed and mobile networks have not been integrated, the handset actually is a dual handset incorporating two terminals, in particular a GSM and a DECT terminal. The dual handset uses the fixed network when in the range of the DECT base station and the GSM network when outside the range. With this type of FMC on the service level, customers still have separate directory numbers for mobile and for fixed services.
- Single directory numbers - customers have only one directory number. This may be a personal number which diverts either to their mobile phone or to the fixed phone where they are at the time. This requires that they call in and let the network know the number where they are located. In this situation, the customer still has both a mobile and a fixed terminal.

Let's now look at the regulatory numbering consequences following from service offers based on one of these elements or a combination of these elements. The first two elements have no consequences for national numbering administration and management. It is the third element that has such consequences. Five cases containing the third element, combined or not with the second element, can be distinguished:

- Personal numbering with separate fixed and mobile terminals - if a service offered only concerns personal numbering then the obvious consequence is that a range for personal numbers should be available in the numbering plan. A personal number range implies that the calling party should be aware of the tariff for a call to a personal number, which in general is higher than the national rate. It is obvious that customers must be able to port their personal numbers both between operators and between service providers.

- Personal numbering with single dual handsets - single directory numbers can be combined with another element in a particular service offer. Of relevance in the numbering context is the combination with single handsets. This combination yields a service offer for which customers have both a single terminal and a single directory number. This can most easily be realized by using a dual handset and a personal number.
- Special numbers for converged fixed-mobile services with single handsets - if the fixed and mobile networks have been integrated ("full" FMC), customers will be able to have a single handset which incorporates only one terminal and they will have a single directory number. The directory number might initially be a number from a range which has been designated for services based upon "full" FMC. A number range for services based upon "full" FMC should be available, the calling parties should be aware of the tariff for a call to such numbers and the numbers must be portable both between operators and between service providers.
- Numbers for mobile and converged fixed-mobile services with single handsets - with "full" FMC, it could be that the numbering plan no longer distinguishes between mobile services and services based upon "full" FMC. This seems in general easy to realize as both types of service use non-geographic numbers without local dialing (which in general is not the case with numbers for fixed local loop services). The directory number could then be chosen from a wide range of numbers that can be used for both types of service. A number range for both mobile services and services based upon "full" FMC must be available, the calling parties should be aware of the tariff for a call to such numbers and NP between all operators and between all service providers offering mobile services or services based upon "full" FMC must be available.
- Number for fixed, mobile and converged fixed-mobile services with single handset - with "full" FMC, it could even be that the numbering plan no longer distinguishes between fixed local loop services, mobile services and services based upon "full" FMC. The directory number could then be chosen from a very wide range of numbers that can be used for any of these services. This case clearly has the most extensive consequences for national numbering administration and management. It is obvious that NP between all operators and between all service providers offering fixed local loop services, mobile services or services based upon "full" FMC must be available [Ber00, Jab92, Kry03c, Iss05].

2.7.3 *Future Interworking*

Already customized application of mobile enhanced logic (CAMEL) is maturing, allowing operators to deliver similar services, such as prepaid, across international boundaries. This level of integration, supported by roaming agreements between operators and the extension of IN functions in the fixed network, enables the personalized communications world to travel with subscribers as they go, or to be fully available on both fixed and mobile phones (Figure 2.14).

The ability of an operator to offer business virtual private network (VPN) services across fixed and mobile networks on an international basis is a powerful weapon in winning high-revenue corporate customers. It can help mobile operators enter the fixed marketplace, existing wireline operators to expand and protect their global markets, or allow entirely new operators to create unique spaces for themselves.

On a more technical level, the introduction of platform-independent solutions such as Java advanced intelligent network (JAIN) supports the development of IN applications able to run seamlessly across a wide variety of fixed and wireless network architectures and protocols. The concept of "write once, run anywhere" is already driving innovation in both services and in new handset and terminal developments.

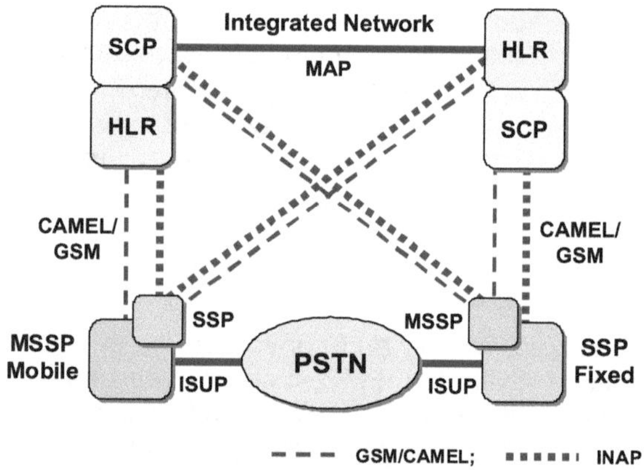

Figure 2.14 - FMC network architecture.

Finally, interaction at a large number of levels between the worlds of internet protocol (IP) and the PSTN will become increasingly important in bridging the fixed and mobile worlds. With agreement now reached on the air interface aspects of UMTS, a greater focus will be applied to the switching issues, with IN playing a vital part. One important role in the converging fixed-mobile environment that IN/IP interaction has to play is in allowing users to create and manage their own service environments through browser interfaces [Bak04, Choi00, Kry03b, Mor00, Telc00].

2.8 Converged IN as a Single Platform for Telephony, Multimedia, and Data Services

The IN platform can be presented as a single platform to offer telephony services regardless of the network access and transport technology used by the carrier and its customers. Instead of maintaining multiple platforms and managing multiple implementations of the same service, a single converged solution can simplify the introduction of new services. Once IN technology is leveraged to offer service convergence, it becomes apparent that the solution space is not limited to traditional voice telephony services. The true utility of IN technology is not so much a function of the PSTN as it is a function of the types of capabilities it can offer to provide voice telephony services. IN platforms perform the following functions:

- Authentication and screening
- Flexible routing and translation
- Storage and access of users profiles
- Flexible charging

These are exactly the same roles that IN could play in offering data and multimedia services. If IN can forward, screen, and charge for a voice call, so it can do these actions for data and video calls, too.

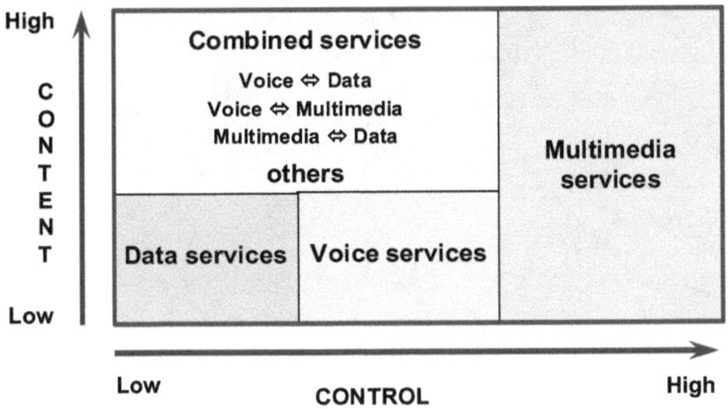

Figure 2.15 - Services classification.

2.8.1 Service Classification

The service scope can be characterized across two different dimensions: content and control. Most IN services deal with call control, such as determining: where, how, when, and if to route a call. The above list of IN functions is used to assist in providing this control. The other dimension of a service is the content, which represents the information conveyed to the users. Telephony services contents provided by the called destination is not really a function of the network. Some telephony services provide content in the form of pre-recorded announcements, but today, the content is a relatively smaller piece of the service. Future multimedia ser-vices, such as "video on demand" or "interactive gaming", have signify-cantly more focus on providing content to users. Figure 2.15 presents a two-dimensional (control and content) mapping of different types of services.

The call control is the main IN functional area, so services in the left part of the diagram would be the best candidates for IN involvement. These services require minimal content to be completely implemented into an IN structure. Services in the right part require high content delivery and high control. They can be offered by a new converged service platform for the service control and the content delivery. However, the IN platform, such as SCP, is not the most efficient delivery vehicle for directly offering multimedia content. This functionality is better left to the domain of specialized servers (e.g., video servers and high speed graphic processsing equipment). But, the network that provides call control and coordination (e.g., IN) can effectively provide interactions with multimedia users and route them to the new IN servers, to offer a complete solution for different types of services support.

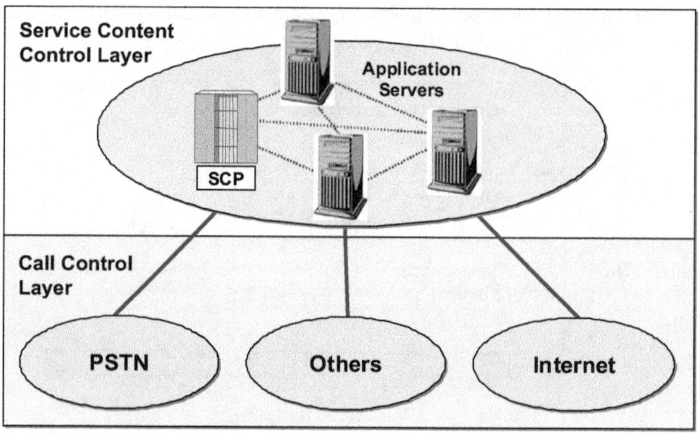

Figure 2.16 - The unified service content and call control model.

2.8.2 *Unified Service Content and Call Control Model*

By evolving the IN platform, the carrier can exploit IN technology for new services offerings while simultaneously achieving the goal of service convergence, as shown in Figure 2.16. The SCP, working with other application servers, can provide services to IP, PSTN, and other networks customers. The service control and content will be provided by the most appropriate system depending on the application, independent of the type of a customer.

The creation of a unified service content and call control model allows to use converged IN as a single platform for telephony, multimedia, and data oriented services. The proposition of a hierarchy of network intelligence is one of the steps to that goal. From our point of view, the model has to consist of two parts: service content control and call control. The call control is responsible for the control dimension while the service content control for the content of services.

Network intelligence is getting now a new meaning. The true intelligence means not only intelligent call control functions, but also service's content control, too. The unified service content and call control model, in that case, allows to establish hierarchical network intelligence, as it is shown in Figure 2.17.

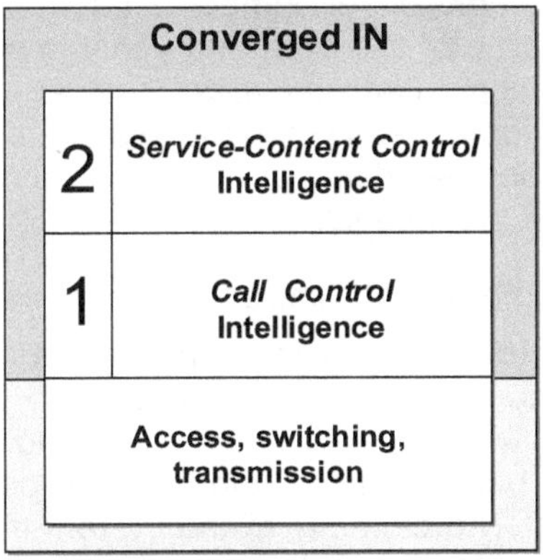

Figure 2.17 - The hierarchy of network intelligence.

That hierarchy presents the layered separation of intelligent functions in converged IN. It allows to concretize future steps for the standardization and creation of a single services platform (based on converged IN). The converged Intelligent Network concept can be viewed as a future seamless platform for circuit-switched and packed-switched networks. The converged IN must be a packet-based network where packet switching and transport elements (e.g., routers, switches, and gateways) are logically and physically separated from the service and call control intelligence. This control intelligence is used to support all types of services over the packed-based transport network, including everything from basic voice telephony services to data, video, multimedia, advanced broadband, and management applications, which can be just another type of service that converged IN supports. A converged IN environment is an important target architecture for a network operator in the near future. How a network operator reaches this goal depends on the services it wants to offer and when it wants to offer them, and IN can play a key role in the process [Kry01, Kry03a, Ram00].

2.9 IN Services Behaviour Modeling

Telecommunication networks exist in a variety of forms and sizes and are used to share information of all kinds and for all reasons. The transfer of this information across networks creates systems with traffic flows of many complexities. The purpose of traffic analysis and simulation is to understand some of the processes that affect the performance of the network. Mathematical analysis and computer simulation are two methods for investtigating network traffic performance. These modeling techniques provide an insight into the mechanisms that operate in complex systems. A mathematical analysis consists of number of equations that express a measure of interest in terms of the fixed and variable parameters of the system. The method of computer simulation involves developing models in software. The simulation is used when the system is complex and the mathematical analysis becomes difficult. The most thorough investigations include a comparison of the results using both techniques.

2.9.1 Performance Analysis of Different Traffic Flow Processes

The traffic presented to a network is often referred to as the offered load. The portion of the offered load that is successfully received at the intended destination is known as the throughput. Carried traffic is another term for throughput that is most often used in the engineering discipline known as teletraffic analysis. The capacity of the system may be measured in a number of ways. One of these is the maximal possible value for the throughput. Often some of the traffic is lost somewhere in the network and the throughput, or carried traffic, is less than the offered load. It takes a certain amount of time for the traffic to be conveyed from source to destination, and this delay is also of major interest.

2.9.1.1 Network Capacity

The communication network capacity should provide an analysis of the traffic quantity with which the system can cope. A number of different parameters can be used to analyse capacity, and the most suitable normally depends on the system under heavy load. A simple example involves calculating the bandwidth of a single link, and in this case it provides an adequate measure of the capacity. However, when it is considered that devices may be connected to each end of the link and that the capacity of these devices may become important, then there are implications for the overall system capacity. The maximum amount of data that may be buffered in the nodes could be a more appropriate measure in this case.

The nature of the traffic may also have an effect on the true value of capacity, and this may introduce time dependency. For example, a telecommunications link may have a capacity for a particular number of simultaneous telephone calls. In certain circumstances, however, the distribution of calls in a switch can adversely affect blocking, and the number of simultaneous connections is then reduced. And, when the system becomes more complex the measure for capacity is less easily defined and more difficult to calculate. Capacity is typically measured in Erlang, bit/s, and packet/s.

2.9.1.2 Network Throughput

The network throughput is a value of how much traffic is successfully received at the intended destination. Hence, the maximum throughput is equivalent to system capacity. Ideally, throughput is the same as the offered load, which is the amount of traffic actually transmitted, and this may be true in cases when the channel is error free, for example. In any case the throughput may only equal to the offered load up to system capacity. The throughput is often less than the offered load when the system is operating below capacity.

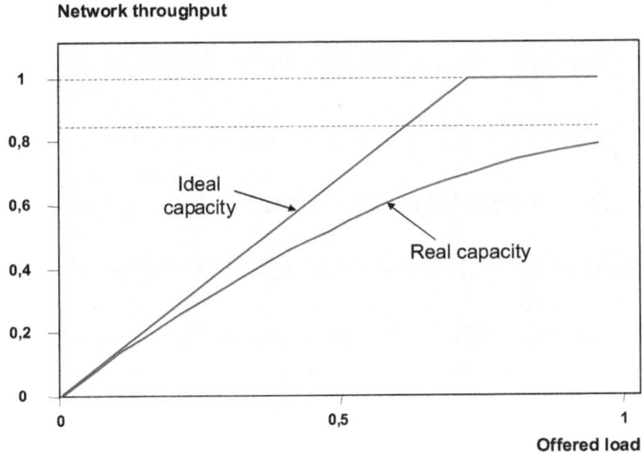

Figure 2.18 - Network throughput for offered load.

The typical network throughput for offered load is depicted in Figure 2.18. The ideal characteristic presents a linear region where the two quantities are equal. The point where offered load and throughput equals one is that at which the system has reached the ideal capacity. This point is indicated by the uppermost dashed line. The actual characteristic is included on the graph to show that the throughput may be less than the offered load at any one point. The actual capacity of the system, indicated by the asymptote to the actual throughput, is therefore less than the ideal capacity. The extent of the departure from the ideal characteristic depends on the efficiency of the communication protocols, blocking probabilities, and so on.

2.9.1.3 Traffic Loss Probability

The traffic loss probability is a value of the chance that traffic is lost. There is a number of situations that result in the loss of traffic. For example, a packet may arrive at a full buffer and it may be involved in a collision, or a call set-up request may arrive at a completely busy switch with no waiting facility. The value of loss probability obtained depends on the traffic intensity and its distribution.

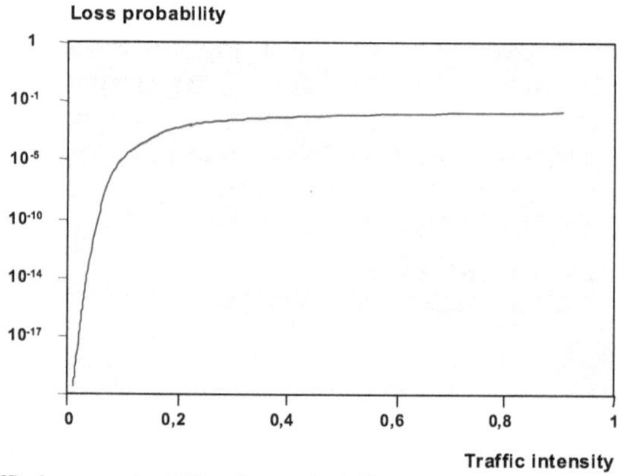

Figure 2.19 - Traffic loss probability characteristic.

Traffic loss probability parameter has traditionally been used as one of two parameters in teletraffic analysis. The second parameter is that of time in system or delay. In the simplest case, loss probability provides an estimate for the likelihood that a call is blocked at a switch. In a more complex situation it may be used, in conjunction with the expected delay, to estimate the performance of a blocking switch with waiting facilities. There are two different measures types of loss probability in teletraffic analysis and these are known as time congestion and call congestion. Time congestion is the portion of time during which all trunks are busy, hence it is the probability that a call arrives to find that it will be blocked. Call congestion is the proportion of calls lost in the long run, and there is a subtle difference between the two quantities. A typical loss probability characteristic is presented in Figure 2.19. The curve shows that the probability of blocking increases with the intensity of traffic and that when the arrival rate is low the loss probability is less significant.

2.9.1.4 The Mean Time in System (Delay)

In the simplest case the delay consists of the time required to transmit the traffic and it is not normally possible to reduce this value. In many cases there are additional sources of delay that contribute to the overall value. Often the transmission delay is insignificant compared with the time required to schedule the communication, or the queueing time, for example. As usual, the value obtained depends on the parameters of the system and the distribution of traffic. Often the parameters can be manipulated in such a way as to reduce the overall delay incurred by the traffic.

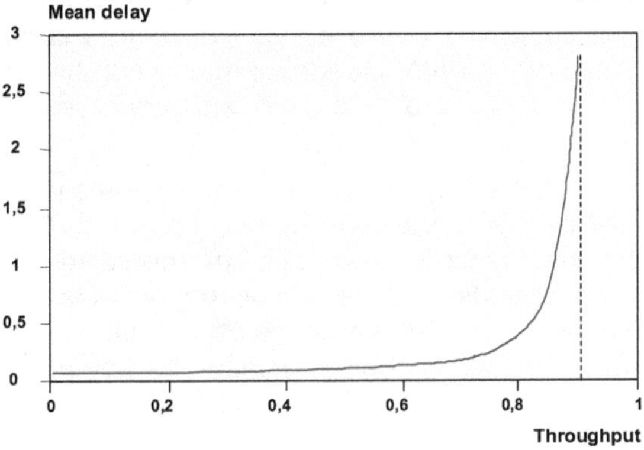

Figure 2.20 - Mean time in system or delay.

The mean delay is one of the traditional measures in teletraffic analysis that is relevant to switches with queueing facilities. The mean delay is also plotted as a function of traffic intensity, throughput, and offered load to provide a performance characteristic that is useful in investigating the efficiency. A typical mean delay versus throughput characteristic is shown in Figure 2.20, wherein the delay is in units of time and the throughput is normalized.

2.9.1.5 Mean Waiting Time or Queue Length

If there are waiting facilities in a communication network, queues will form at points of congestion. Often the length of a queue may be a parameter of interest. For example, if the mean queue length, or the average number of packets waiting to be switched at an input, can be predicted, then this may be used to estimate the required length of a buffer. It is likely that the nature of traffic in the system will result in variations from the mean that are unpredictable. Moreover, the length of a buffer has an effect on other parameters of interest, such as the loss probability. In general, the queue length is directly proportional to delay. Thus a typical queue length characteristic is similar to that presented in Figure 2.20 with delay on the vertical axis that can be replaced by queue length in units of packets.

2.9.1.6 Queueing Theory Telecommunication Networks

The analysis of communication network traffic involves statistical mathematics. The buffers in any communication network form queues of data waiting to be transmitted. It is reasonable to expect that the amount of data in a queue will have some effect on how long it will take before the data may be transmitted. It is also reasonable to expect that the occupancy of the queue will be determined by a number of factors such as the arrival and departure rates of the data. Queueing theory can be applied to the investtigation of these factors to determine the properties of different systems.

The complete performance investigation of all systems will normally involve calculations of two or more performance parameters. A thorough exploration into the mechanisms of the system should reveal the nature of trade-offs between various parameters of the network and its traffic. A performance investigation normally involves a qualitative comparison of so-called performance characteristics. These characteristics are obtained by drawing the graphs that show how a performance measure changes when one or more of the system parameters are varied. When it is the case that the analysis is straightforward, a mathematical model in the form of a simple equation can be obtained. This equation will express the desired performance value in terms of the parameters of the system. These parameters include quantities such as the number of nodes and communicating devices, the packet and call generation rates, the lengths of messages, the size of buffers, and so on. The complexity of the system is normally the factor that dictates whether such an equation is obtainable. It is often possible to simplify the model by making assumptions about the parameters of the system. Of course, it may be the case that the investtigation is to account for the effects of these parameters and we are again faced with the problem that their inclusion as a variable in the model is a factor that complicates the analysis. One of the most common assumptions required to simplify the analysis is that the system is in steady state. In this case all the performance measures obtained are average values and the effects of any variations in time on the measure are not provided.

Analysis that do account for the time-varying effects, so-called transient solutions, is however sometimes possible. The other common assumption is that there are an infinite number of sources for traffic. In this case, when the arrivals are also assumed to be completely random they may be characterized by the Poisson and exponential distributions, which are of-ten most convenient. In any case the assumptions should be clearly sta-ted in order to preserve the ability to keep the results of an investigation in context.

And, finally, there are two approaches to mathematical analysis. One is to use first principles in deriving equations that are directly relevant to the system under scrutiny. The fundamental results of probability theory are often useful in this case. The other is to apply the results of queueing theory, which is a well-established branch of mathematics, and use standard techniques.

2.9.1.7 Simulation of Telecommunication Networks

The computer simulations are an alternative or supplement to mathematical analysis. Often two methods are put to use in conjunction with each other to provide a process of validation. Comparing the results of analysis and simulation is a useful and rewarding exercise when the system under scrutiny is complex. Often it is possible to verify the correct operation of a simulation using an analysis of the equivalent system, each with the same set of assumptions. The analysis alone may not permit such a study. Given sufficient computational resources, and the time in which to utilize them, it will always be possible to produce a computer simulation for a network of any degree of complexity. Hence, the method of computer simulation may be called upon to investigate the systems that cannot be dealt with using analysis. The choices for simulation are numerous. There is a number of obvious considerations that may be highlightted, and these are applicable to each of the preceding options. First and foremost the resources that are accessible will provide the options that are immediately available. Next, a clear idea of the objectives for the investtigation is required. After this, it is impossible to define a structured approach that is generally applicable to any investigation. There is, however, a number of characteristics that may be considered important. Flexibility is one of these. For example, if a model is developed, then how easy could it be modified. User-friendliness is another. If the model is developed by one person, then how easy would it be for another to use it. The amount of development time required is probably also of concern, and the tools available for analyzing the results may be another. Developing simulations, from first principles using a high-level language; permits the user complete control over the definition and operation of the network to be simulated. The method is popularly known as that of discrete event simulation. This is due to the nature of the program in its quest to replicate a sequence of defined actions that occur at both regular and irregular points in time. In particular, the method of pointer variable manipulation is appropriate for the task of handling event lists. Also, increasing process-sing speed is one of the factors that should enhance the lot of a computer simulator. In particular, when so-called rare-event processes are of inte-rest the run-time for a simulation can be excessive [Hig98, Pat97].

2.9.2 Performance Modeling of Telecommunication Networks

For the analysis of a telecommunication system, a model must be set up to describe the whole or parts of the system. This modeling process is fundamental especially for new applications of the teletraffic theory; it requires knowledge of both the technical system as well as the mathematical tools and the implementation of the model on a computer. Such a model contains three main elements (Figure 2.21):

- the system structure,
- the operational strategy, and
- the statistical properties of the traffic.

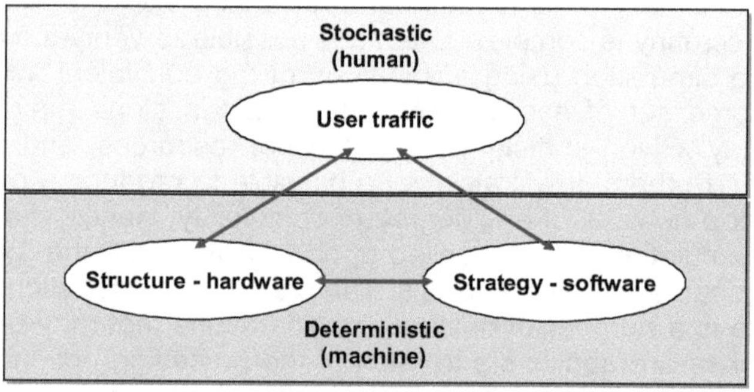

Figure 2.21 - Telecommunication systems model.

2.9.2.1 System Structure

This part is technically determined and it is in principle possible to obtain any level of details in the description e.g. at component level. The reliability aspects are stochastic and will be considered as traffic with a high priority. The system structure is given by the physical or logical system which normally is presented in manuals.

2.9.2.2 Operational Strategy

A given physical system can be used in different ways in order to adapt the traffic system to the demand. In a computer, this adaptation takes place by means of the operation system and by operator interference. In a telecommunication system strategies are applied in order to give priority to call attempts and in order to route the traffic to the destination. In stored program control (SPC) telephone exchanges, the tasks assigned to the central processor are divided into classes with different priorities. The highest priority is given to accepted calls followed by new call attempts whereas routine control of equipment has lower priority. The classical telephone systems used wired logic in order to introduce strategies while in modern systems it is done by software, enabling more flexible and adaptive strategies.

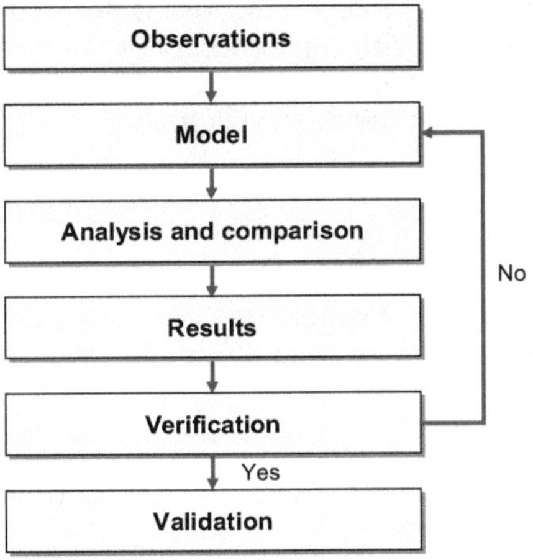

Figure 2.22 - Modeling process.

2.9.2.3Statistical Properties of Traffic

User demands are modeled by statistical properties of the traffic. Only by measurements on real systems is it possible to validate that the theoreticcal modeling is in agreement with reality. This process must necessarily be of an iterative nature (Figure 2.22). A mathematical model is build up from a solid knowledge of the traffic. Properties are then derived from the model and compared to measured data. If they are not in satisfactory accordance with each other, a new iteration of the process must take place.

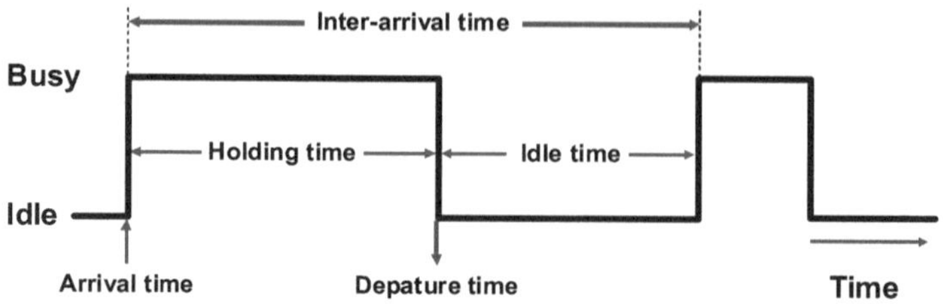

Figure 2.23 - Traffic process

It appears natural to split the description of the traffic properties into stochastic processes for arrival of call attempts and processes describing service (holding) times. These two processes are normally assumed to be mutually independent meaning that the duration of a call is independent of the time the call arrived. Models also exists for describing users experiencing blocking, i.e. they are refused service and may make a new call attempt a little later (repeated call attempts). Figure 2.23 illustrates the terminology usually applied in the teletraffic theory.

2.9.2.4 Models

General requirements to a model are:

- It must without major difficulty be possible to verify the model and it must be possible to determine the model parameters from observed data.
- It must be feasible to apply the model for practical dimensioning.

The variations observed in the number of ongoing established calls in a telephone exchange, which vary incessantly due to calls being established and terminated. Even though common habits, daily variations follow a predictable pattern for the subscriber behavior, it is impossible to predict individual call attempt or duration of individual calls. In the description, it is therefore necessary to use statistical methods. The call attempt events takes place according to a stochastic process, and the inter arrival time between call attempts is described by those probability distributions which characterizes the stochastic process [Ive01].

2.9.3 Finite Source Models

The M/M/1/K/K queueing system has been studied since the 1950's when it was used to study the performance of weaving machines that would break after random intervals, given that these breakages would need to be fixed by an operator who tended several machines. This model was developed in the mid 1970's and applied to computer performance modeling. It found applications in the modeling of processes waiting for the fulfilling of I/O requests those requests coming from several independent processes. It was also used in the modeling of user queries against some database where the operating time is the time taken by the user to generate their next query. This abstract model is often referred to in queueing theory literature as the machine repair with one repairman or the machine interference model (Figure 2.24).

It can be seen that for a process making requests of a distributed virtual memory system there is a correspondence with this model. The operating time is the time that the processor is performing useful work between requests for data, while the broken time is the time spent waiting for a response from the virtual memory system.

There are several important performance parameters that can be derived from this model. Some of these assume exponential distribution for both the service time and the time for which the processor runs before requesting service, whereas others only require the conservation of flow principle and hence hold for other request and service time distributions.

Table 2.1 summarizes these parameters and highlights the difference in physical interpretation between the modeling of parallel computing systems and the more usual machine repair model.

Parameter	Virtual memory performance model	Machine repair
$1/\lambda$	Average time between requests for data	Average time between breakdowns, the operating time
T_S	Time taken to service a particular request	Time taken to repair a particular breakage
$1/\mu$	Average time to service data request	Average time to repair a machine
q	Number of requests queueing awaiting service	Number of broken machines awaiting repair
N	Number of requests outstanding (number of processors that are idle awaiting data)	Number of machines that are down
λ	Average arrival rate of requests from the parallel system	Average rate at which the K machines break down

Table 2.1 - Comparison of M/M/1/K/K interpretations.

It can be seen that the performance of the system corresponds to assigning meaning to the duration of the location of the tokens in various locations within the representation. The same sort of technique has been used in various other settings the probabilistic duration calculus maps the state of the system to either 1 or 0. The integration of this variable can then be used to qualify an attribute of this system over the phases of its operation. In its application to computer performance the model has been used to find bounds on the number of users of transaction processing systems; to estimate the maximum number of sources that can share packet switching network connections; and to examine the response times for multiple users of a CPU system.

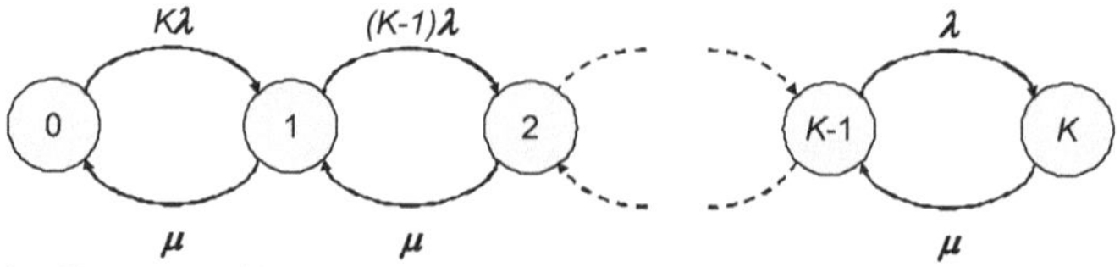

Figure 2.24 - State-transition diagram for M/M/1/K/K system.

2.9.3.1 General Properties

The general properties of this system could be derived equally well by reference either to the structural model or to the behavioural model combined with the conservation of flow.

The relationship between the request rate, service rate, server loading and wait time can be derived using only the conservation of flow:

- Let the utilization of the server be ρ. The server is assumed to be eager, i.e. it processes a request whenever there is one in the queue to be processed.
- The system throughput, T_S, is defined as the average number of requests that are processed per unit time; this equals the number of requests that arrive from the processors per unit time by the conservation of flow. As the network processes μ requests (on average) per unit time while the server is busy then:

$$T_S = \rho\mu \tag{2.1}$$

- Letting the response time from the server be T (this includes the time spent queueing for service and the time spent receiving service), the average interval between successive requests from each processor in the system is $T+1/\lambda$. This implies that the average rate of requests from each processor is $1/(T+1/\lambda)$.
- Since there are K processors, all running independently, the total arrival rate of requests, which equals the throughput of the system T_S is

$$T_S = \frac{K}{T + 1/\lambda} \tag{2.2}$$

By the conservation of flow, equations (2.2) and (2.1) must be equal. Solving these equations for the average response time T, gives

$$T = \frac{K}{T_S} - \frac{1}{\lambda} = \frac{K}{\rho\mu} - \frac{1}{\lambda} \tag{2.3}$$

As already stated this argument does not rely upon the distribution of either the service times or the request inter-arrival times. The resulting equation (2.3) and other results derived from it are valid for any queueing system in which the average service time is $1/\mu$ and the average interval between requests is $1/\lambda$.

2.9.3.2 Distribution of Related Properties

The relationship between the load intensity (u) and server loading (ρ) is dependent upon both the request and service rate being exponential distributions.

The length of the queue in Table 2.1 can be used to represent the state of the system. The system that is represented by these states can be viewed as a Markov birth-death system. In such a system all the running processors are generating requests independently of each other. As illustrated in Figure 2.24, the following transitions between the states can occur:

- state s_j to s_{j-1} - represents a request being serviced, the re-starting of an idle processor and a reduction in the number of items to be serviced.
- state s_j to s_{j+1} - represents a new request being made by an active processor, that processor becoming idle, and an increase in the number of items to be serviced.

In this model a processor can generate a request only while it is active and, once a request has been made, it cannot generate another request until that request has been serviced. Thus the state number, which is equivalent to the number of items queued, represents the current number of processors that are idle. When the system is in state s_0 all K processors are running.

First, looking at the instantaneous transition rate from state s_j to s_{j+1}, the probability that any one processor will make a request in a time interval h is $\lambda h + o(h)$. Also, given that s_j represents j idle processors, there must be $K-j$ processors running and hence possible candidates to generate the next request. Thus in s_j this set of processors which are running independently of each other has an overall probability that a request will be made of $(K-j)\lambda h + o(h)$. This gives an instantaneous transition rate from s_j to s_{j+1} of $(K-j)\lambda$. The instantaneous transition rate from state s_j to s_{j-1} does not change with the state of the system, as there is only one service point and this is operating at a rate of μ. Denoting the steady-state probability of state s_j by p_j we have the following set of local balance equations:

$$K\lambda p_0 = \mu p_1$$
$$(K-1)\lambda p_1 = \mu p_2$$
$$\vdots$$
$$\lambda p_{K-1} = \mu p_K$$

which is equivalent to:

$$(K-j+1)\lambda p_{j-1} = \mu p_j; \quad 1 \le j \le K \tag{2.4}$$

From these equations it can be seen that it is possible to express the probabilities of all the states in terms of p_0, namely:

$$p_j = \frac{K!}{(K-j)!} u^j p_0; \quad 1 \le j \le K \tag{2.5}$$

Where,

$$u = \lambda / \mu \tag{2.6}$$

The normalization equation

$$\sum_{i=0}^{K} p_i = 1$$

implies that the value of p_0 is:

$$p_0 = \left[\sum_{i=0}^{K} \frac{K!}{(K-i)!} \left(\frac{\lambda_r}{\mu} \right)^i \right]^{-1} \tag{2.7}$$

2.9.3.3Performance Measures of this Model

- Server Loading - The server has work to perform whenever any one of the processors has made a request, i.e. whenever the system is not in state s_0. This gives the server fractional utilization ρ as:

$$\rho = 1 - p_0 \tag{2.8}$$

Another useful measure is the total rate of requests from all the processsors. This measure λ must equal the rate at which the server is processsing them, thus:

$$\lambda = \rho\mu = (1 - p_0)\mu \tag{2.9}$$

This value λ can be viewed as the message density that the server is processsing.

- Average Queue Length and Response Time - The average response time T has already been defined in equation (2.3) and by Little's result the average length of the queue for processors requesting and receiving service is:

$$\begin{aligned} L &= \lambda T \\ &= \rho\mu T \\ &= K - \frac{\rho\mu}{\lambda} \\ &= K - \frac{\rho}{u} \end{aligned} \tag{2.10}$$

where, $\rho = 1\text{-}p_0$ from equation (2.7).

The total response time T consists of two components: the time spent in the queue waiting for service T_q and the time spent receiving service T_s. T_s is the average service time - $E[s]$, which equals $1/\mu$.

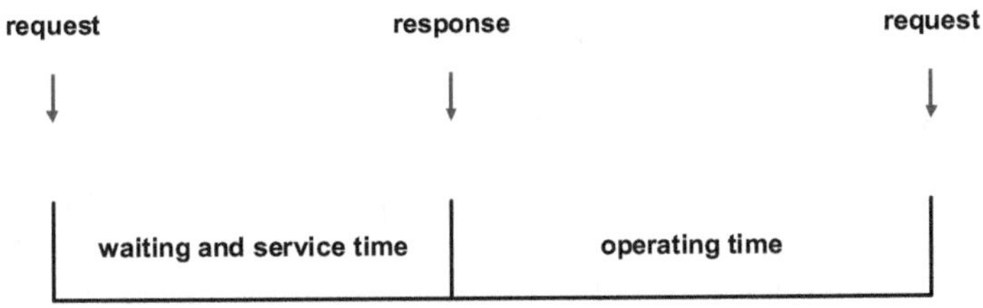

Figure 2.25 - A request, cycle of a single processor.

Thus the time spent queueing is:

$$\begin{aligned} T_q &= T - T_S \\ &= T - E[s] \\ &= \frac{K}{\rho\mu} - \frac{1}{\lambda} - \frac{1}{\mu} \end{aligned} \tag{2.11}$$

- Processor Idleness / Processor Utilization - As there are no other overheads in this model, a processor is idle whenever it is waiting for a response to a request; this is T as derived in equation (2.3). When a processor is running it generates requests at the rate λ; each processor cycles between the two states as illustrated in Figure 2.25.

To complete one of these cycles takes $T + 1/\lambda$ period of time. Hence the idleness of this processor is:

$$Fraction - Idle = \frac{T}{T + 1/\lambda} \tag{2.12}$$

If the equation (2.12) is rewritten to find *T* in terms of the fractional idle, and if it is combined with the definition of *T* from equation (2.3), an equation for *u* using the measure of fractional idle can be derived:

$$\begin{aligned} \frac{1}{\lambda}\left(\frac{idle}{1-idle}\right) &= \frac{K}{\rho\mu} - \frac{1}{\lambda} \\ u &= \frac{\rho}{K(1-idle)} \end{aligned} \tag{2.13}$$

This provides for a measure of the loading intensity *u* based on the measurement of the per-processor idle time. However, when using the definition of ρ from equation (2.8), a polynomial of order *K* is generated; for which a direct algebraic solution is too complex in the general case. Such polynomials are amenable to numerical solutions. In the case *K=1* it simplifies to

$$u = \frac{idle}{1-idle} \tag{2.14}$$

2.9.3.4 Practical Use of this Model

Although straightforward and not complex, the results of the queueing system that describes this model have many potential uses. Within the limitations of the original model, those being the absence of delay and overhead, the qualitative results hold for any distribution of request and service rates where use of an average is reasonable. In the use of quantitative results there is a need for the user of this model to have justification that the exponential distribution assumptions are reasonable for the particular application.

Exponential distributions (and the renewal Poisson processes derived from them) have many physical manifestations, and many physical processes are adequately modeled by them. Even if a single random process is not well approximated by a Poisson process, there is justification that, as the output from several random processes are superimposed, the resulting random process can be approximated more and more accurately by a Poisson process.

In a particular system it may not be the software of the virtual memory server that is the subject of contention but the communications between processors. The Poisson process has been successfully used for the modeling of communication behavior. The usual justification is that the data being transmitted is of varying length and the transmission time is related to the length of the message being transmitted. If the subject of contention is some software server then the major portion of the service time is the execution of the algorithm that implements the service. Although most searching algorithms have a uniformly distributed execution time, the use of caching and the underlying scheduling of processes will perturb this.

If it is felt that the distribution of request and service times is not reasonnably modeled by Poisson processes, the formulae that are derived by substituting $1\text{-}p_0$ for the value of ρ would need to be avoided. This still allows for the comparison of various measures but does not allow for the predictive aspects, as these need to predict the value of ρ for a number of processors. Where distribution assumptions are felt to be reasonable it is only necessary to know u in order to predict the various performance measures of the system. This can either be derived by some direct measurement, analysis of the code itself, or indirectly.

2.9.3.5 Model Shortcomings

This model assumes that a processor waits for a response to every request before continuing useful work; this waiting can constitute a signifycant portion of the run time. It assumes that there are no delays in moving data around the system.

It assumes that μ and λ remain constant as the configuration changes; this is not always so. The request rate λ depends on many factors which change with the number of processors and with topologies.

Finally, the model takes no account of other overheads such as CPU costs for initiating requests, the costs of routing messages, or other work such as load balancing [Dav94, Yad98].

2.9.4 Intelligent Network Analysis by Closed Queueing Model

With increasing deployment of intelligent network services, design and engineering of network intelligence platforms to accommodate the ever-changing and growing demands of customers, presents a rich market of opportunities and challenges, although tempered by concerns arising from the problematic experiences of similar system and network developments. As the telecommunications industry evolves, customers are increasingly coming expecting instantaneous access to service providers, together with transparency to network failures. System performance dictates that response times need to be minimized, sufficient redundant capacity to be installed in case of failure and controls embedded within the design to manage the exceptional situations that continually threaten network integrity. The service scenario mixes service demand, physical network topology, signaling message flows, the mapping of functional entities to physical components, and routing as part of the network design process to ensure that performance requirements are met [Ack97, Kol98].

For the system performance aspect, the most significant changes due to intelligent network (IN) are the distribution of network intelligence and the new services made possible by this distributed architecture. Whereas traditionally a call is processed within the switch, in the IN environment, a call involves the cooperative processing of several network elements connected by a signaling network. This fundamental change possesses some new challenges to teletraffic experts to ensure that IN networks are designed to provide customers' services with good performance [Yan94].

There is the necessity to have network performance and capacity figures already available in the service specification phase. It is also important to properly consider the IN as an integrated part of a telecommunication network. The additional load generated by these new IN services may lead to a performance degradation that can spread beyond the IN environment, which, in turn, affects not only the quality of the new IN services, but also the services already offered. In our work, the modeling approach is based on the construction of model for the various components of the IN architecture leading to a multiple-chain queuing network system. The analysis is conducted using hierarchical decomposition techniques, allowing a detailed consideration of the signaling network protocol [Baf96].

2.9.4.1 IN Architectural Concept

The IN is an architectural concept that provides for the real-time execution of network services customer applications in a distributed environment consisting of interconnected computers and switching systems. The rationale for it is the provisioning of a multitude of supplementary services (such as call forwarding, freephone, conference calling, e.g.), which require the development of service logic along with basic call processing [Lic01].

The IN concept, which in the third phase is known as IN capability set 3, is currently subject to international standardization effort and a set of recommendations have already been made available. A framework for the design and description of the IN architecture is given by the IN conceptual model. This model is divided in four planes addressing service aspects, global functionality, distributed functionality and physical aspects of an IN.

The service plane represents an exclusively service-oriented view and no aspects regarding the implementation of the services are handled. The services of the service plane are mapped onto the service logic in the global functional plane. Service logic is a chained sequence of standardized, monolithic, and reusable network capabilities called service independent building blocks (SIBs).

The functionality required to support the realization of the SIBs is contained in the functional entities of the distributed functional plane. These functions include end user access and interactions, service invocation and control, and service management. The interface between the user and network call control is the call control agent function (CCAF). The call/service processing and control is provided by the call control function (CCF). The real-time call processing service logic is contained in the service control function (SCF). The service switching function (SSF) is the set of functions required for interaction between CCF and SCF, e.g., recognition of service control triggers, signaling management, etc. The customer and network data are stored in the service data function (SDF). These functional entities are mapped onto the physical entities of the physical plane [Baf96].

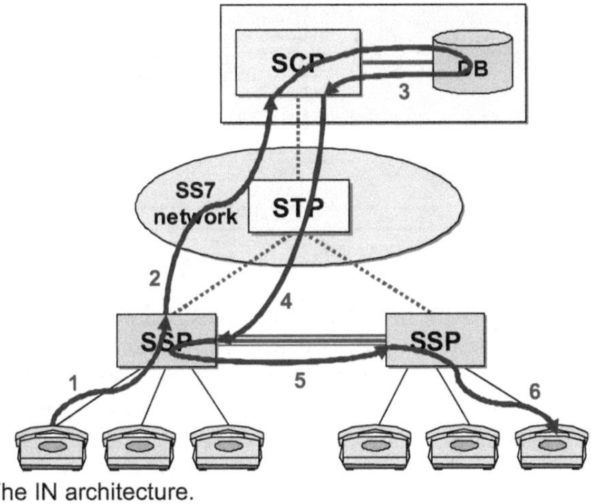

Figure 2.26 - The IN architecture.

The schematic representation of the IN structure is presented in Figure 2.26. As shown in the figure, the IN consists of service switching points (SSP) that accommodate terminals, service control point (SCP) that store databases (DB) and perform advanced service control, and the signaling system No.7 (SS7) network that transfers messages between SCPs and SSPs via signal transfer points (STP).

An IN service typically involves interactions between the SSPs and some of the other IN nodes to perform call and connection control, interact with the user, and monitor events on signaling interfaces. For example, the SSP may temporarily suspend call processing to send a query to the SCP where SCP-based services, such as user authentication, number translation, route selection, alternate billing, etc., are executed. The SSP may pass control to the intelligent peripheral (IP) to get more information from the user [Kry02, Kry04, Yan94, Mas01].

2.9.4.2 Closed Network as Intelligent Network Model

A queueing network is a collection of service stages (devices, in a telecom system) in which customers proceed from one stage to another to satisfy their service requirements. The network is closed if the number of customers in the system remains fixed.

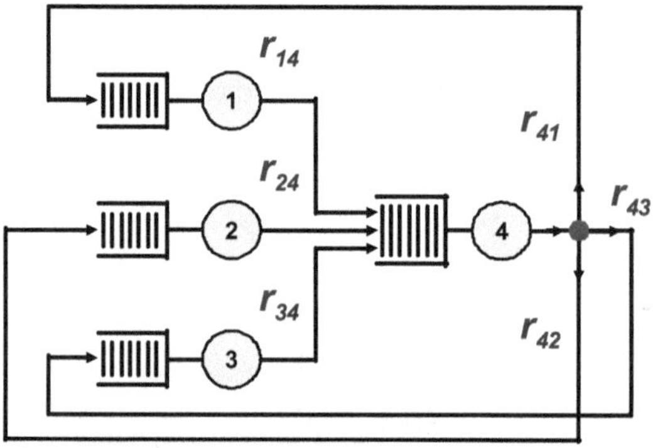

Figure 2.27 - The closed queuing network.

A closed queueing network is shown in Figure 2.27. The circles are separate service devices (servers), and the attached squares indicate spaces for queues of customers awaiting service at the server. The arrows indicate possible paths between servers. When more than one arrow leaves a server, a customer may go to one of several other servers after completing service at that one. At least one arrow must leave each device. A job moves instantaneously from one device to another; it enters ser-vice instantaneously upon arrival at an idle server; and, on arrival at a busy server, it joins the queue of customers awaiting service and remains at that server until completing its service requirement there. The state of the system, at any given time, is a vector of the number of customers present at each of the services. For example, in the system schematized in Figure 2.27, if at some instant of time there are 3, 0, 1, and 4 customers at servers 1, 2, 3, and 4 respectively, then the four-tuple (3, 0, 1, 4) descry-bes the state of the system. For this state the number of customers in the system is fixed, at 8. One customer is in service and two customers are awaiting service at server 1, server 2 is idle, one customer is in service and no customers are waiting at server 3, and so on.

The customers spend random lengths of time in service at the servers and travel random paths around the system. Therefore, only a probabilistic descripttion of the system's behavior can be given; this description quantifies the chance that the system is in a given state at a given time. The description specifies the probabilities of observing various values of service times and the probability that a customer leaves a server by a particular path. In Figure 2.27, the quantity $r_{i,j}$ denotes the probability that a customer goes to server *j* upon completing a service at server *i*. If there is no (*i, j*) arrow, the probability $r_{i,j}$ is zero. When arrow (*i, j*) is the only one from server *i*, then $r_{i,j} = 1$ since a customer leaving server *i* is certain to follow that path. It must always be true that the $r_{i,j}$ sum to 1 over all *j*, since it is certain that a customer leaving server *i* will go to some other server. For server 1 in the example of Figure 2.27, $r_{i,j}$ is not zero because it is possible for a customer to return to server 1 for further service at the end of a time slice. The service-time probabilities are expressed by functions that give the probability that the random service time T_S, of a customer at server *i* does not exceed a given length of time *t*. For server *i*, this function is denoted by

$$f_{S_i}(t) = p(T_{S_i} \leq t), \qquad t \geq 0; \tag{2.15}$$

It is called the service-time probability distribution function. Using the r and the *f(t)* to obtain an overall description of the random process may present extremely difficult theoretical and computational problems unless some simplifying assumptions (approximations) are adopted. The simplest such assumption is that the service-time probability distribution at each device is an exponential distribution, i.e.,

$$f_{S_i}(t) = 1 - \exp(-\mu_i t), \qquad t \geq 0; \tag{2.16}$$

where $1/\mu_i > 0$ is the mean or average service time at server *i*. A second simplifying assumption is that it is possible for a customer at any server *i* to proceed eventually to any other server *j*, possibly requiring intermediate stops at other servers. In other words, it is impossible for a customer to remain forever in some isolated part of the system. Under these two assumptions, the process of system states changing over time can be modeled by an "irreducible Markov chain".

A useful property of these chains is memoryless conditions: at some future time the distribution of probabilities of the various system states is approximately the same regardless of the initial conditions. As time goes on, these probabilities converge to limits called steady-state probabilities. These steady-state probabilities have a useful practical interpretation: they can be interpreted as the long-run proportions of time the system spends in the various states. In particular, the (long-run) proportion of time for each possible queue size at each server can be calculated, which in turn allows the calculation of the (long-run) proportion of time that each server is in use.

For closed queueing networks with exponentially distributed service times, Gordon and Newell derived the steady-state distribution of the vector of queue sizes. Buzen showed how to compute the steady-state distribution efficiently and in particular how to calculate the proportion of time that a given device is busy. Some of these results are briefly described in the next section of our paper [Lip77].

2.9.4.3 Queueing Network Models Usage

Queueing network models are widely used in communication and computer systems to represent the flows in a system. Basically, there are two types of queueing networks: the open queueing network and the closed queueing network. The open queueing network provides service for arriving customers from an external source, and the completed customers depart from the network. For closed queueing networks, no external source of customers exists, while a fixed number of customers circulate indefinitely among the servers in the network [Dia01].

The IN can be presented as a network of queues where the total number of customers (e.g., SSPs) is fixed because no customers are allowed to arrive or depart, which is closed. Closed networks can be analyzed using Markov chains. And, the steady-state occupancy distribution has a product form under assumptions similar to those used for open networks.

It is taken into the consideration a network of K queues, where:

- There are no external sources for customers and there no external destinations for customers;
- There are $K < \infty$ customers in the network;
- There is a single server at each queue. The service times at the *i-th* queue are independent exponential random variables with rate μ_i. service times at different queues are independent;
- The routing through the network is random. The probability of a customer leaving queue i for queue j is $r_{i,j}$.

This network is called a *Jackson closed network*. The state of a *Jackson closed network* is described by

$$\underline{n} = (n_1, n_2, \ldots, n_i, \ldots, n_K), \tag{2.17}$$

where n_i is the number of customers at queue i. The steady-state probability of being in state n is once again denoted by $p(n)$.

The global balance equation of the *Jackson closed network* can be set up according to the balance between the net flow (total rate) into the state and the net flow (total rate) out of the state.

$$\sum_{i=1}^{K} \mu_i p(\underline{n}) = \sum_{i=1}^{K} \sum_{j=1}^{K} \mu_i r_{i,j} p(\underline{n} - \underline{1}_j + \underline{1}_i), \tag{2.18}$$

which is called the global balance equation at state $\underline{n}$.

In terms of the traffic equations, we can solve the global balance equations to obtain the following product form solution

$$p(\underline{n}) = p(\underline{0}) \prod_{i=1}^{K} \left(\frac{\theta_i}{\mu_i} \right)^{n_i}, \tag{2.19}$$

where

$$p(\underline{0}) = \left\{ \sum_{\underline{n}} \left[\prod_{i=1}^{K} \left(\frac{\theta_i}{\mu_i} \right)^{n_i} \right] \right\}^{-1}, \tag{2.20}$$

Let θ_i be the average throughput through queue *i*. Then,

$$\theta_i = \sum_{j=1}^{K} r_{j,i} \theta_j, \qquad i = 1, 2, \ldots, K, \tag{2.21}$$

which are the traffic equations of the network [Ber92].

2.9.4.4 M/G/1/K/K Queueing System Application to the IN Architecture

Traditional PSTN services were provided by the service logic and data resident within the local switching machine. The capacity for these services is very much determined by the architecture and component capacities within the service node. IN has a distributed architecture in which the service logic is executed cooperatively by different network elements that can be geographically dispersed [Yan94].

Figure 2.29 shows an IN network distributed architecture as a finite source model - M/G/1/K/K, also known as the machine repair model, or the cyclic queue model. In that context, there are *K* requests cycling in a system consisting of *K* SSPs and a central processor unit (CPU) with a work queue representing the SCP. A request for IN service is sent from the SSP to the SCP after an exponentially distributed think time and after being processed by the SCP the response returns to the SSP which enters in another think phase. The input and output messages of a transaction are treated as a single composite service. Also, the think time and the SCP processing time are considered as an average operating time [Hoo83].

The M/G/1/K/K system always reaches the steady state (Figure 2.28) because there can be no more than *K* requests in the system (one request being served, and *K-1* waiting for serving or are in thinking time). The service time in this system has a general distribution although the up time for each SSP is exponential [All90].

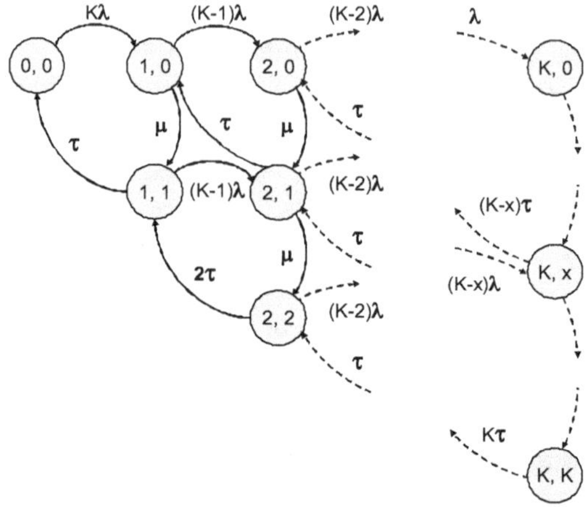

Figure 2.28 - State-transition diagram for M/G/1/K/K system.

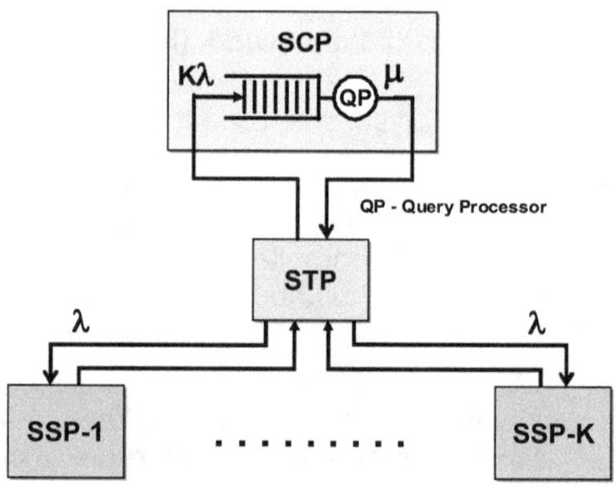

Figure 2.29 - The IN distributed architecture as M/G/1/K/K model.

The rate at which requests enter the SCP is equal the rate at which responses leave, because the system is assumed to be in equilibrium. The period of time that a particular SSP is in the thinking state is merely the ratio of the average thinking time $1/\lambda$ to the average time $T + 1/\lambda$ which it spends in making a complete round trip. Each of the *K* SSPs generates requests at a rate λ per second, provided it is in the thinking state. On the other side, the period of time when the SCP is busy is $1-p_0$, during which time the output rate is μ. Thus the average output rate of jobs is $\mu(1-p_0)$. Equating the input and output rates we have:

$$K\lambda \frac{1/\lambda}{T+1/\lambda} = \mu(1-p_0) \tag{2.22}$$

From that equation, we obtain the equation for *T* [Kle86a, Zep94, Kle86b]:

$$T = \frac{K/\mu}{1-p_0} - \frac{1}{\lambda} \tag{2.23}$$

The numerical results for the expected response time, which is the sum of the waiting time in the queue and the service time (e.g., average time in system - *T*) are given in Figure 2.30. In Figure 2.30, we have presented *T* for several values of *K*. Here the service time is exponentially distributed [Hoo83].

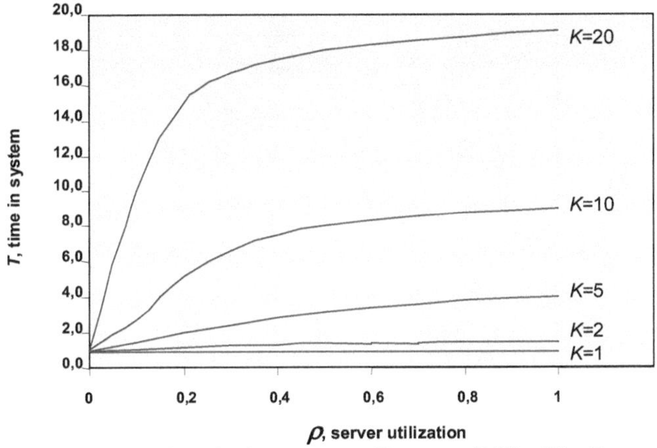

Figure 2.30 - The average time in the system - *T*, versus SCP utilization.

In Figure 2.31 is plotted the normalized response time as a function of the number of SSPs, *K* for different values of SCP utilization. In this figure, we can see very slow rise in the response time as the number of SSPs increases from 1. But, after passing through a transition area, normalized response function becomes linear. This behavior is easily explainable - in area, where the response time is growing very slowly, it is clear that the number of requests from SSPs is so small that the periods when a customer needs service are usually the periods when other customers are thinking and therefore not interfering with him. But, when the number of requests increases, the normalized response time become linear since p_0 approaches zero [Kry02, Kry04, Kle86a].

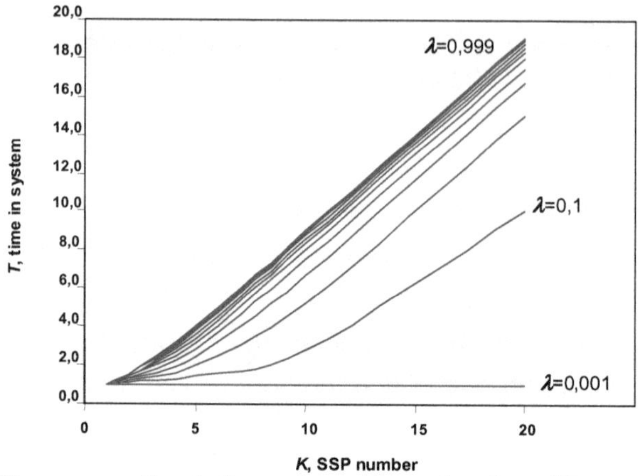

Figure 2.31 - The average time in the system - T, as a function of the number of SSPs.

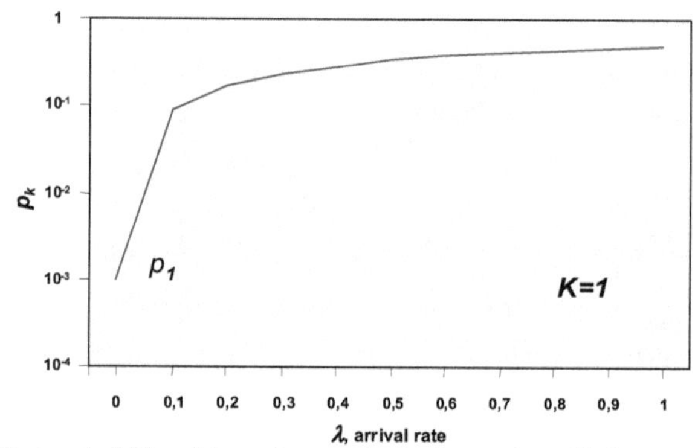

Figure 2.32 - The probability of k customers in system, when K=1.

The M/G/1/K/K is a system, where number of customers is finite, namely K. And, it behaves in the following way - a customer is either in the system, waiting or being served, or outside the system and arriving. The interval from the time he leaves the system until he returns once again is exponentially distributed with mean $1/\lambda$. This case gives the following equation for the probability for finding k customers in system:

$$p_k = \frac{K!}{(K-k)!}\left(\frac{\lambda}{\mu}\right)^k p_0, \qquad 0 \le k \le K \tag{2.24}$$

where, p_0 – is, as usual, the probability that there are no customers in the system:

$$p_0 = \left[\sum_{k=0}^{K} \frac{K!}{(K-k)!} \left(\frac{\lambda}{\mu} \right)^k \right]^{-1} \qquad (2.25)$$

The results of the probabilities for finding *k* (*K* = 1, 2, 5, 10, 20) and 0 customers in system, are given in Figures 2.32 ÷ 2.37.

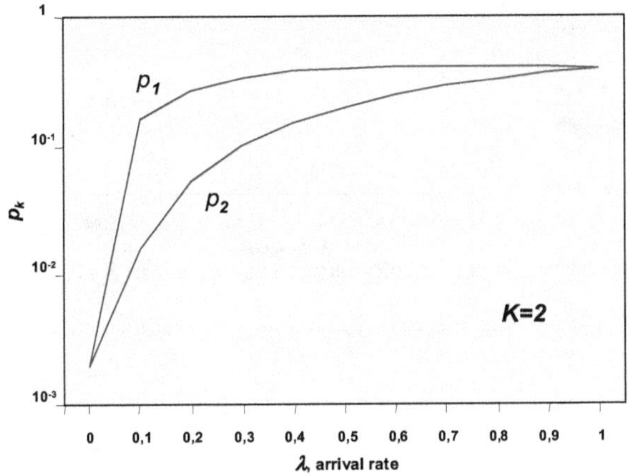

Figure 2.33 - The probability of *k* customers in system, when *K*=2.

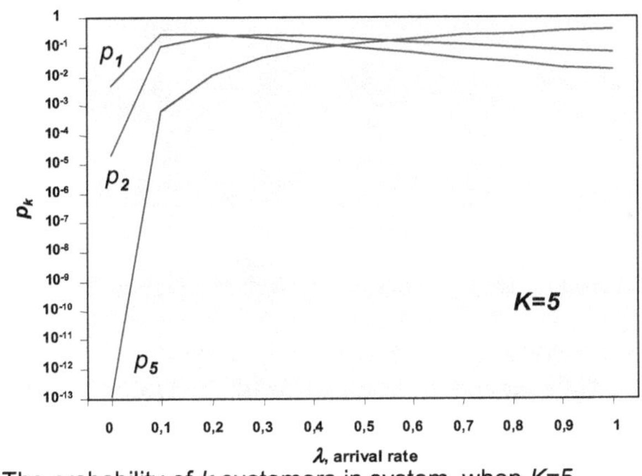

Figure 2.34 - The probability of *k* customers in system, when *K*=5.

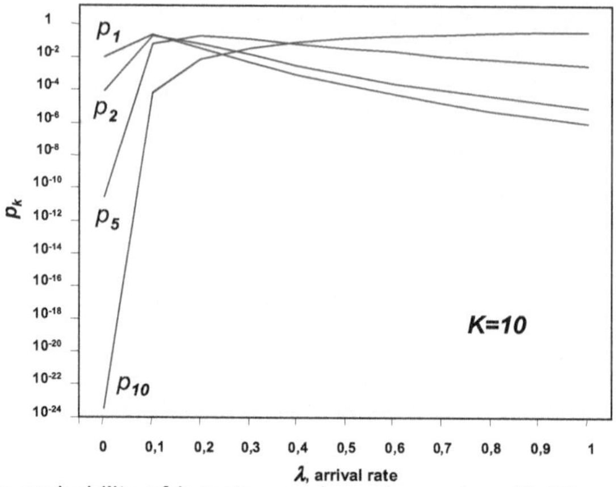

Figure 2.35 - The probability of *k* customers in system, when *K*=10.

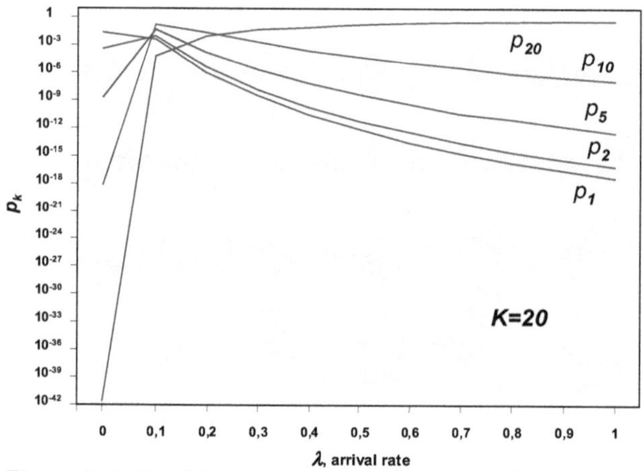

Figure 2.36 - The probability of *k* customers in system, when *K*=20.

In Figure 2.38 we show the average time in system T for different coefficients of variation of service time for M/G/1/K/K system. The variance of a random variable was defined as

$$\mathrm{var} = \sigma^2 = \frac{1}{n}\sum_{i=1}^{i=n}(x_i - mean)^2 \qquad \Pr(X = x_i) \tag{2.26}$$

Mostly, in queueing theory is used squared coefficient of variation of service time C_S^2.

$$C_S^2 = \frac{\sigma_{T_S}^2}{T_S^2} = \frac{\text{var}}{mean^2}. \tag{2.27}$$

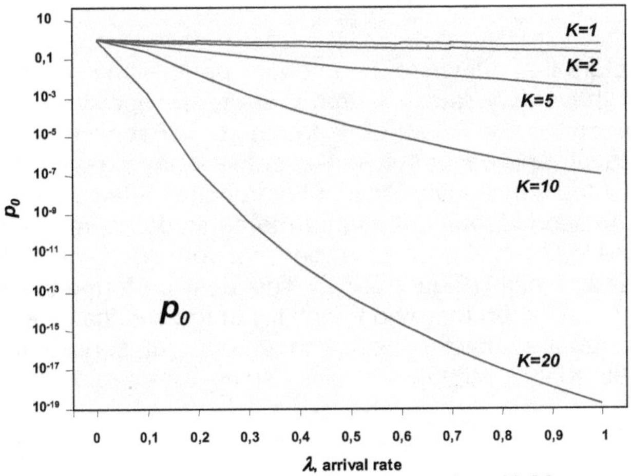

Figure 2.37 - The probability of *0* customers in system, when *K*=20.

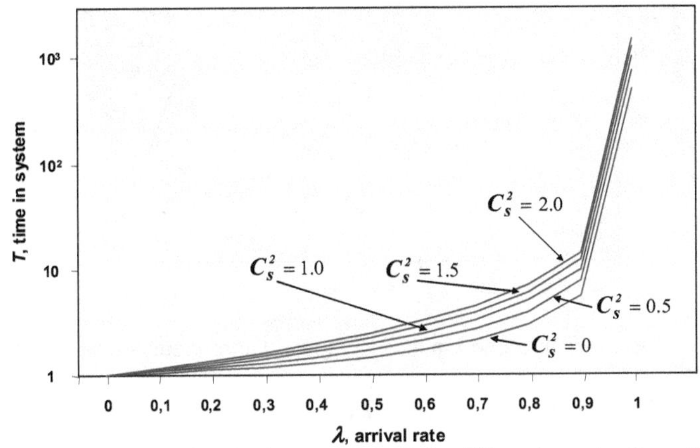

Figure 2.38 - The average time in the system - *T*, for different service time distributions

The average time in system *T* is computed for the following service distributions:

- D – Deterministic service time (C_S^2 = 0);
- E_2 – Erlang-2 service time (C_S^2 = 0.5);
- M – Exponential service time (C_S^2 = 1.0);

- H – Hyperexponential service time (C_S^2 = 1.5) and (C_S^2 = 2.0).

$$T = \left(1 + \frac{\rho\left(1 + C_S^2\right)}{2(1-\rho)}\right) T_S \tag{2.28}$$

The parameters that influence the average performance of the M/G/1/K/K system are the arrival rate - λ, the average service time - T_S, and the second moment or coefficient of variation - C_S^2 of service time. When the server utilization approaches 1, the average delay grows without bound. The shape of the service time distribution function affects performance only through the second moment, higher moments do not matter. Even a not heavy loaded M/G/1/K/K queue can perform very poorly when the variability of demand is high (Figure 2.39). The best performance is achieved when C_S^2 = 0, i.e. when there no variability in the service time, all requests from SSPs require exactly the same amount of service time. [Kry02, Kry04, Tan95, Kihl97, Mit98].

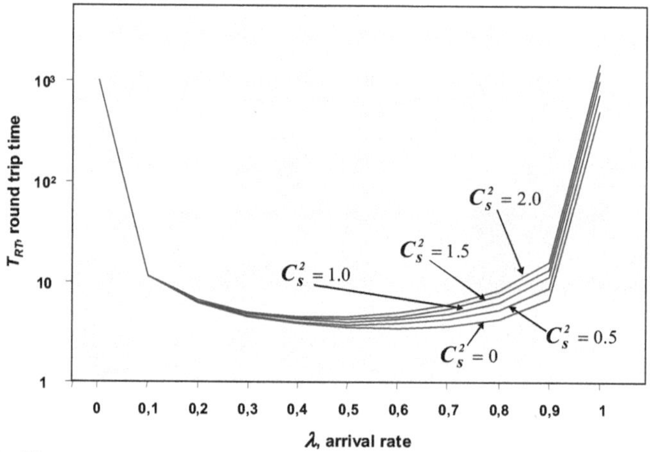

Figure 2.39 - The average round trip time – T_{RT} for different service time distributions.

2.9.5 *M/M/2/K/K System for Large IN*

2.9.5.1 Distributed SCP

In intelligent networks, the terminal location, authentication data, and other information used when dialing the number that has been assigned to a terminal are stored in a DB at an SCP. The demand for IN services, existing and new ones, continues to grow. This means that SCPs have to accommodate very-large-scale DBs or they have to have distributed architecture (Figure 2.40).

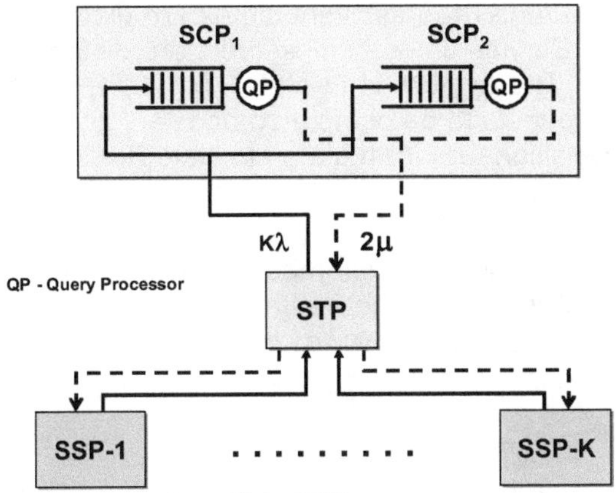

Figure 2.40 - Large scale IN with multiple SCPs

There are two basic strategies that SCPs could be made:

- multiple SCPs - deployed as independent node on the network each with its own DB;
- single distributed control SCP with multiple modules has the ability to interconnect multiple modules that contain a distributed DB.

From the standpoint of both simple implementation and maintenance of IN, it is desirable that SSPs can access SCPs without being aware of the distributed SCP topology. This means that, when an SSP sends a request to an SCP via the SS7 network, it is not necessary for the switch to select and specify the address of certain SCP. This can be achieved in one of two ways:

- access request from the SS7 network is multicast to all SCP;
- access request from the network is sent via any SS7 link to the target SCP.

In the scheme 1, nothing needs to be done to achieve distributions invisibility inside SCP, while in scheme 2, it will be apparent that an invisibility condition exists. That is, an access requests received from the SS7 network do not specify the target SCP, so an additional condition is required in order to access the target SCP [Mas01, Kry03a].

2.9.5.2 M/M/m/K/K Model for Large IN Modeling

M/M/m/K/K is presented a general model of a system with K users and m parallel servers. The model is illustrated in Figure 2.40. There are at least as many users as servers. If $K < m$, then $m - K$ servers are never used and may be discarded. The user think times are distributed exponentially with parameter λ. Service times at all servers are distributed exponentially with parameter μ. The system is in state j (j = 0, 1, ..., K) if j users are waiting for their requests to be completed and K - j users are thinking. The instantaneous transition rate from state j to state j+1 is equal to

$$\lambda_j = (K - j)\lambda, \qquad j = 0, 1, ..., K - 1; \tag{2.29}$$

since each of the thinking users submits requests at rate λ. The rates from state j to state j-1 depend on whether the number of requests is less than the number of servers, in a similar way to M/M/m/K/K model:

$$\mu_j = \begin{cases} j\mu & \text{for} \quad j = 1, 2, ..., m - 1 \\ m\mu & \text{for} \quad j = m, m + 1, K. \end{cases} \tag{2.30}$$

The state diagram is shown in Figure 2.41.

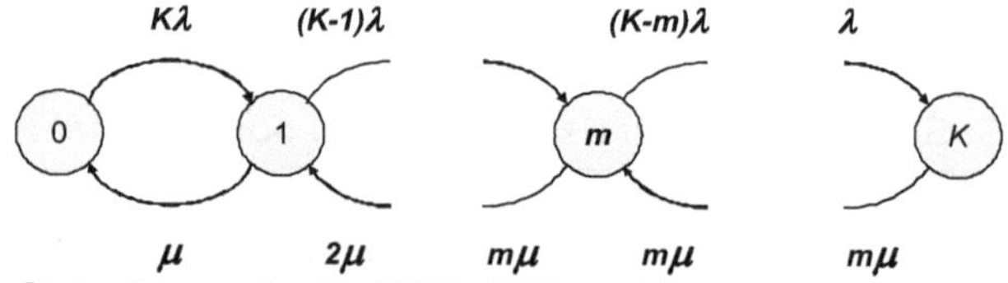

Figure 2.41 - State diagram for the M/M/m/K/K model.

The balance and normalizing equations yield

$$p_j = \frac{K!}{(K - j)!\,j!}\rho^j p_0; \qquad j = 0, 1, ..., m - 1$$

$$p_j = \frac{K!}{(K - j)!\,m!\,m^{j-n}}\rho^j p_0; \quad j = m, m + 1, ..., K, \tag{2.31}$$

with p_0 given by

$$p_0 = \left[\sum_{j=0}^{m-1}\frac{K!}{(K - j)!\,j!}\rho^j + \sum_{j=m}^{K}\frac{K!}{(K - j)!\,m!\,m^{j-m}}\rho^j\right]^{-1} \tag{2.32}$$

The throughput T, can be obtained either as the average number of requests completions, or as the average number of requests submissions, per unit time. The former approach requires the average number of busy servers, r:

$$r = \sum_{j=1}^{m-1} jp_j + m\sum_{j=m}^{K} p_j \tag{2.33}$$

The expression for the throughput is then $T = r\mu$. Alternatively, we could find the average number of requests in service or in the queue, L:

$$L = \sum_{j=1}^{K} j p_j \tag{2.34}$$

Then the average number of thinking users is K-L. Since each of them submits requests at rate λ, the throughput is equal to $T = (K-L)\lambda$.

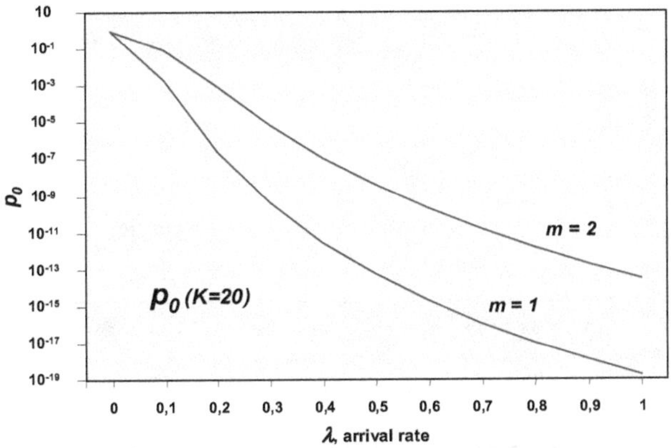

Figure 2.42 - The probability of *0* customers in system (K=20) for m =1 and 2 servers.

In the two special cases when $m = 1$ and $m = K$, the expressions have a simpler form. If there is a single server, the steady-state probabilities are

$$p_j = \frac{\rho^j}{(K-j)!}\left[\sum_{i=0}^{K} \frac{\rho^i}{(K-i)!}\right]^{-1}; \qquad j = 0, 1, \ldots, K \tag{2.35}$$

and, the throughput is equal to $T = (1-p_0)\mu$.

When the number of server is equal to the number of users, no request has to queue and users do not interfere with each other in any way. The steady-state distribution of the number of requests in service is binomial:

$$p_j = \frac{\rho^j}{(K-j)!}\left[\sum_{i=0}^{K} \frac{\rho^i}{(K-j)!}\right]^{-1}; \qquad j = 0, 1, \ldots, K \tag{2.36}$$

The average number of busy servers is $r = K\rho/(1+\rho)$. The throughput is given by [Mit98]:

$$T = \frac{K\lambda}{1+\rho} \tag{2.37}$$

In Figure 2.42, it is shown comparison of the probabilities that there are 0 customers in system (K=20) for m =1 and 2 servers in system. And, as it was expected, the probability that there are no customers in system, of course, is higher for m=2.

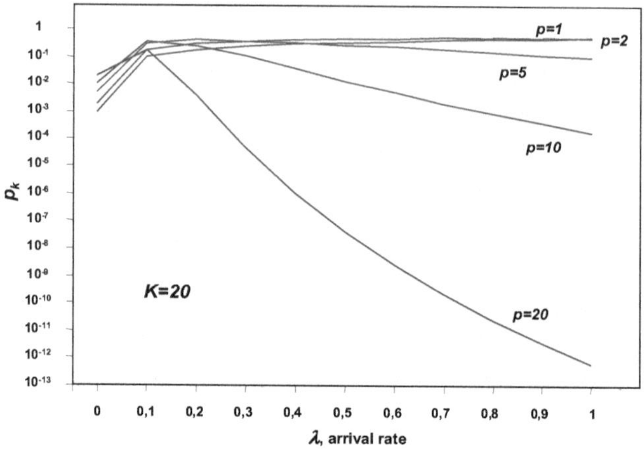

Figure 2.43 - The probability of k customers in system (K=20; m=2).

In Figure 2.43 are given results of the probabilities for finding k (K = 1-20, m=2) in system. In comparison with m=1 server system, the probability of k customers in system, when m=2 servers is decreasing, because service rate is growing.

The average number of customers in system (e.g., in the queue or service) L is plotted in Figure 2.44. In Figure 2.44 we also show the average time in system T, when m=2 servers, and distribution of the service time is exponential.

The analysis and modeling of system performance issues are essential tools in the development and engineering processes that can be used at all stages of the lifecycle of IN services. Simple, approximate models have a high value in the early stages to uncover major performance problems which affect the design of the architecture before the cost of rectification is too high. The design tools support rapid prototyping, allowing users to go through the three important stages: predict, design, and comparison. The questions of the development of new modeling methods for rapid analysis, and some others, like new IN performance standards, and closer connections between performance analysis and service design are the most interesting for the future wide deployment of IN services [Kry03a].

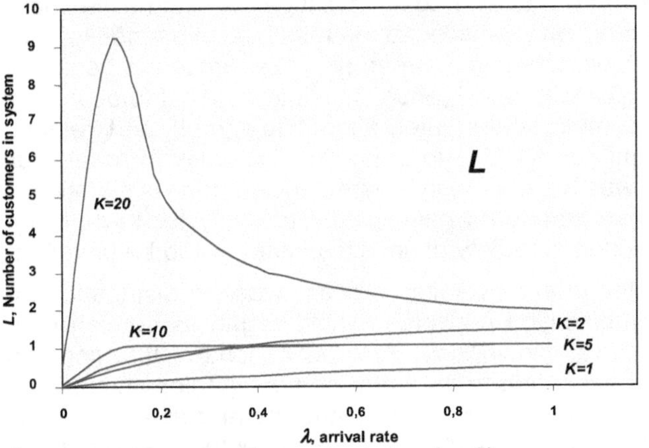

Figure 2.44 - Average number of customers in system (*m*=2).

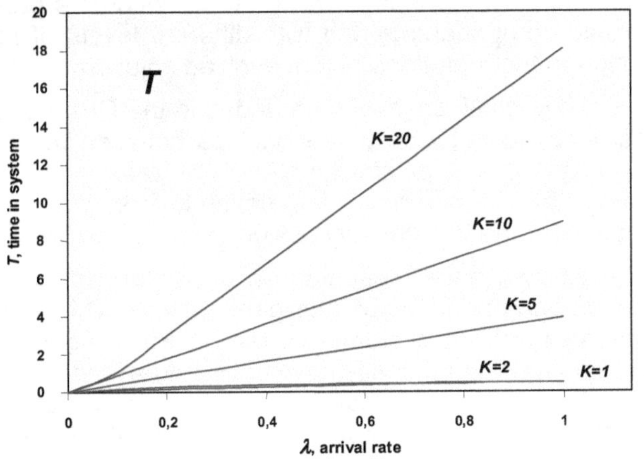

Figure 2.45 - The average time in system *T* (*m*=2).

2.10 Chapter Contributions Summarizing

In this chapter, network intelligence evolution process has been presented. It started from the simplest conceptual telephone networks, which were the result of years of evolution, with little thought about future technologies. Based almost on analog equipment, telephone networks were not able to support such services as data and video. Many individual technology service providers began popping up providing packet-based networks and data communication services the telephone companies were just not equipped to provide.

As international bodies began investigating alternative technologies for providing telephone services to the mass, such as mobile, the need for an all-digital network became apparent. Thus, arose the beginnings of an all-digital network with intelligence. The international telecommunication union (ITU) commissioned international telegraph and telephone consultative committee (CCITT) to study the possibility of an all-digital network. The result was a series of standards known now as signaling system no.7 (SS7). These standards have paved the way for the intelligent network (IN), and, with it, a variety of services, many yet to be unveiled.

The next step in this evolution process was the distribution of intelligence in telecommunication networks, which began as nothing more than segmentation of responsibilities. The foundations of that segmentation were established according to the trend of moving telecommunication solutions toward more diverse computing platforms and away from monolithic settings. With movement and diversity has come the ability to integrate new solutions into the overall base system with greater speed and efficiency. Ultimately, the base system has transformed to become a part of a larger set of integrated components, each with different levels of responsibility and contribution to the intentions of that evolved solution.

Distributed network intelligence was grounded in the OSI model. It is within this model that the necessary relationships between multiple hosts of the distributed network are established. The OSI reference model defines a partitioning of network operability into seven layers, where one or more protocols implement the functionality assigned to a given layer.

In the section 2.6, we have presented an example of platform that supports distribution and openness of intelligent network. This section descrybes a service for handling customers calling a travel agency. It is assumed that the service was originally developed and available for PSTN only. It is shown how the service can be extended to include interfaces to additional access networks like mobile networks and the Internet.

The convergence between fixed and mobile networks, which until recently were two separated worlds, is next logical step of networks and services intelligence evolution towards single, open platform. The idea to run one single network instead of two separate, one for fixed services and one for mobile services, allows network operators to save money in terms of infrastructure and operational costs. The ability to provide integrated fixed and mobile services offers opportunities both for the provision of promising services and for a recognisable differentiation from competitors. From customers' point of view fixed/mobile convergence gives an integ-rated service package, whereby the customer is offered both types of ser-vices, using one terminal and possibly one number, and receives one bill. It meets customers' expectations in terms of independence from terminals and technologies.

Besides, converged IN, as a single platform for voice, multimedia, and data, can offer these services regardless of the network access and transport technologies used by the carrier and its customers. Instead of maintaining multiple platforms and managing multiple implementations of the same service, a single converged solution can simplify the introduction of new services. Once IN technology is leveraged to offer service convergence, it becomes apparent that the solution space is not limited to traditional voice telephony services. The true utility of IN technology is not so much a function of the PSTN as it is a function of the types of capabilities it can offer to provide voice telephony services. The creation of a unified service content and call control model allows to use converged IN as a single platform for telephony, multimedia, and data oriented services. The true intelligence means not only intelligent call control functions, but also service's content control, too. The unified service content and call control model, in that case, allows to establish hierarchical network intelligence. And, this hierarchy separate intelligent functions in converged IN.

And next, the IN services behaviour modeling involves the analysis of intelligent network signaling traffic using statistical mathematics. The buffers in any communication network form queues of requests waiting to be served. It is reasonable to expect that the amount of data in a queue will have some effect on how long it will take before the data will be served. It is also reasonable to expect that the occupancy of the queue will be determined by a number of factors such as the arrival and departure rates of the data. In this chapter we apply queueing theory methods for the investigation of these factors to determine and calculate IN performance parameters.

The information transfer across networks creates systems with traffic flows of many complexities. The purpose of traffic analysis and simula-tions is to understand some of the processes that affect the performance of the network. In section 2.9.1, we define traffic parameters that need to be analyzed and investigated. Also, we introduce and explain modeling techniques for the investigation of signaling messages flows in telecommunication networks.

In Section 2.9.2, we dealt with the modeling processes for the analysis of telecommunication systems. It has been set up a model that describes the parts or the whole system. And, we have given definitions and explanations of main elements of this model.

The finite source queueing models have been presented in Section 2.9.3. They need to be explained in detail because of use in the next sections of this chapter. It has been shown their evolutional process. There were introduced their general properties. And, performance parameters to be measured were analyzed. Also, the model shortcomings and practical usage examples have been given.

The Sections 2.9.4 and 2.9.5 dealt with queueing models for performance analysis of intelligent networks. For the system performance aspect, the most significant changes due to intelligent network (IN) are the distribution of network intelligence and the new services made possible by this distributed architecture. Whereas traditionally a call is processed within the switch, in the IN environment, a call involves the cooperative proce-ssing of several network elements connected by a signaling network. This fundamental change possesses some new challenges to teletraffic ex-perts to ensure that IN networks are designed to provide customers' ser-vices with good performance.

The additional load generated by these new IN services may lead to a performance degradation that can spread beyond the IN environment, which, in turn, affects not only the quality of the new IN services, but also the services already offered. In our work, the modeling approach has been based on the construction of model for the various components of the IN architecture leading to a multiple-chain queuing network system. The analysis has been conducted using hierarchical decomposition techniques, allowing a detailed consideration of the signaling network protocol.

The converged IN together with its unified service content and call control model has been a final phase of the network intelligence evolution. But, it has been evolution process of pure telecommunication networks, in order new services development and support. Now, telecommunication networks are migrating toward Internet technology. Theherefore, the next chapter concentrates on different integration aspects of IN and IP networks convergence. It gives comprehensive description of all standardizetion and research efforts in this direction.

3 Service Integration Aspects by the IN and IP Networks Convergence

The convergence of the public switched telephone network (PSTN) and Internet protocol (IP) data networks promises exciting opportunities for local and long-distance wireline and wireless carriers, Internet services providers (ISPs), equipment manufacturers, and value-added resellers (VARs). An important step in the fulfillment of this promise is the extension of intelligent network (IN) capabilities to and from IP networks. Telephony IN services such as toll-free service, local number portability (LNP), and calling-card services are now being complemented by a corresponding set of IP-based call management and intelligent routing services. Signaling system no.7 (SS7) and SIP (PINT/SPIRITS) are means by which these applications can achieve interoperability. This integration of IN and IP technologies represents a significant breakthrough in the ongoing convergence of voice and data networks [3Com05, IEC-IN/IP, Wei98, Van99].

3.1 Integration Drivers and Enablers

Telecommunications networks are migrating towards Internet technology, with voice over IP maturing rapidly.

One of the hottest topics in telecommunications today is the use of packet data networks, and the Internet in particular, to transport voice and fax, traffic that classically runs on the circuit-switched public telephone network (PSTN) [Rin99, Schn00].

At the dawn of the third millennium, IP data networks have become a worldwide proliferated communication medium. Today, voice is already being transported over various data networks. However, this is just the start. Media Gateways and controlling Call Servers are about to be introduced, so that the data network infrastructure will not only be able to carry basic voice calls. The full range of value-added voice services complemented by innovative multimedia services is the challenging goal.

3.1.1 Objectives

The benefit of today's rather scare voice over IP applications is primarily cost reduction. Therefore voice is, first of all, carried over the long-distance data network infrastructure. An example is voice over IP interconnect private networks. Considering the extraordinary price-cuttings for telecommunications over the past few years, it is very unlikely that in future, price reduction remains the only argument in a competitive telecommunications market. What is urgently needed is improved functionality for new revenue making services.

3.1.2 *Enablers and New Perspectives*

Progress in voice and video coding in conjunction with emerging of powerful, low-cost digital signal processors (DSP) are surely the necessary prerequisites. Meanwhile, protocol stacks for call setup and media control in packetized networks are also available. However, from the technical point of view the reference architecture for migration, as illustrated in Figure 3.1, indicates that the pure inband media conversion in the media gateway, like voice from the switched-circuit network to voice over IP, is only one part of interworking. The pivotal point for bringing new sophisticated communication services to data networks is a media gateway controller. A rather promising interface between media gateway and media gateway controller is shortly called MEGACO or H.248. Establishment and control of calls in IP data networks furthermore requires reliable transport of signaling information. The internet engineering task force (IETF) currently defines a generic protocol on top of UDP for the transport of signaling protocols like SS7, DSS1 or ISUP.

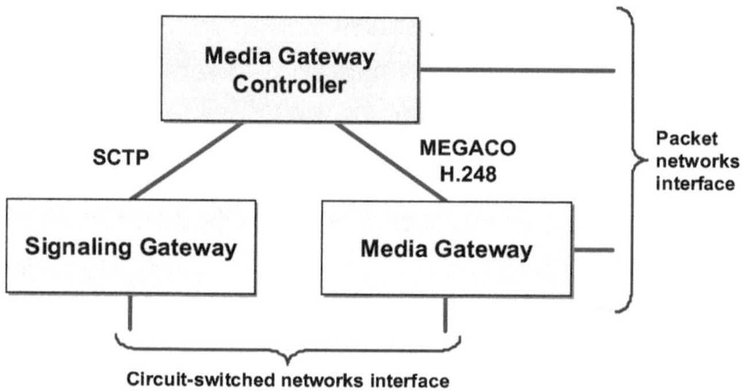

SCTP – Simple Control Transmission Protocol
MEGACO – IETF working group

Figure 3.1 - Gateway Reference Architecture.

It is called simple control transmission protocol - SCTP (Figure 3.1). An important objective of the migration is to carry the intelligence of the existing switched-circuit networks to data packet networks. This can effi-ciently be done by a media gateway controlling call server, which makes ample use of the already proven circuit switched call-handling function-nality. Packet networks compared with circuit-switched networks are ca-pable to offer more flexibility in bandwidth allocation and management. This is an excellent prerequisite for efficient introduction of innovative multimedia communication [IBM05].

3.1.3 Network Configuration and Major Feature Requirements

The key for providing circuit-switched network services through IP is handling different types of access networks. The solutions of today only support pure IP at the customer premises. The challenge is to provide access, which can also handle PSTN and ISDN lines. In legacy circuit-switched networks the above function is supported by the local exchange. Since new operators do not want to install traditional local exchanges, a device that can be connected to the access equipment is required to handle voice and IP traffic. In addition, functions of interconnection to other operators have to be provided. This device can be used either by incumbent local exchange carriers (ILECs) to replace their existing local exchanges, or by competing LECs to install new equipment.

3.1.4 Interfaces and Protocols

On top of the well-known Internet transport protocols various protocols specific for IP telephony need to be supported. Examples include: gateway control protocol, call signaling, control protocol for multimedia communication, and the bearer independent call control protocol.

3.1.5 Prototype

A prototype has been used to demonstrate the network concept. The prototype includes existing access network devices which are capable to terminate ADSL, PSTN, and ISDN lines. The call server software has been derived from software of a public exchange. In this software the V.5 device handlers are mediated in an adaptation layer to control physical devices (the access gateways) via the media gateway control protocol MGCP. The first version of the access gateway/call server prototype can be characterized by:

- Start of a high-performance, future safe platform (e.g., high port density access gateway);
- Support of narrow-band PC internet dial-in;
- Support of ISDN-to-ISDN telephone calls over an IP backbone;
- Support of supplementary services like number identification.

3.1.6 Evolutional Increase of Intelligence to IP networks

A low cost entry into the voice over IP market may be voice trunking. This scenario may in addition support H.323 terminals. Introduction of new and innovative services can make ample use of functions, which have already proven reliability in communication networks. Examples include:

- extended gatekeeper functions;

- service management center functions;
- call server functions;
- extended intelligent network functions;
- service creation environment functions.

Rapid creation of new services furthermore may be opened to competitive third parties. This opens the door for a wealth of imaginative new services and applications.

3.1.7 Outlook

Nowadays, IP telephony is still not in a final phase. VoIP gateways are beginning to be deployed for backbone replacement of incumbent telephone operators and for new competing operators. With the demonstration of the prototype it can be shown, that the demarcation line between circuit-switched and packet-oriented networks can be moved closer to the access. The multiple interworking and gateway capabilities of the access gateway/ call server combination allows for a smooth transition from today's PSTN to an IP network including next generation telephony provisioning [Sch00, Jan00, Sij00].

3.2 IN/IP Convergence - Standardization and Research Work

In recent years, there has been a huge increase in the penetration of desktop computing in homes and businesses worldwide. This has to a large extent been promoted by the success of the Internet, which has clearly demonstrated the immense commercial potential of multimedia communications services. This has raised customer expectations of service features that should be offered by the public telecommunications infrastructure, principally that they support a mix of media types and allow easy customization. Additionally, users expect that these services will be available on demand, regardless of their location or the capabilities of their terminal equipment. While the main technological components required to realize this vision are available, there remains a significant challenge in deploying them in a manner that both is cost effective and can continue to meet the demands of a volatile marketplace.

As the level of interconnection between fixed, mobile, Internet, and enterprise networks increases, a key component in ensuring their ongoing success will be the availability of a common platform for the development and delivery of communications services. Of course, a key requirement for operators who intend to enhance their service delivery capabilities is that existing systems be leveraged as much as possible rather than replaced. Many see the intelligent network (IN), which is today the prevalent means of providing services based on manipulation of voice call setup, as a starting point for the service delivery platform of the future. Currently a number of groups are proposing short to medium-term evolutionary paths for IN aimed at overcoming limitations of existing systems.

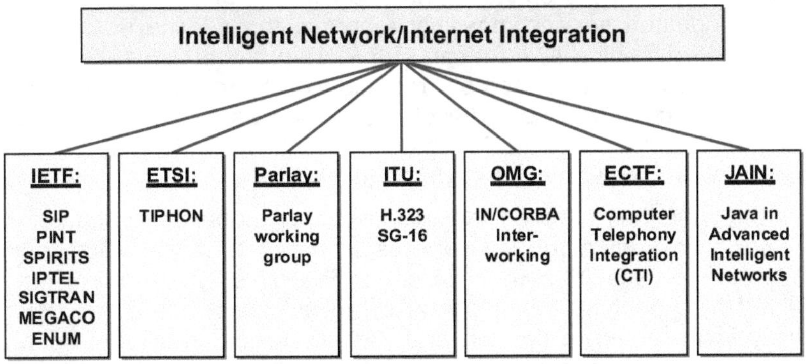

Figure 3.2 - International and industrial bodies developing standards and architectures.

As shown in Figure 3.2, the Internet/information technology and public switched telephone network (PSTN) integration is being carried out by several standardization bodies [Bre00, Cha00, Den03].

3.2.1 PSTN/Internet Interworking (PINT)

The PINT working group is part of the IETF transport area, and was created in 1997 to address how Internet applications can request and enrich telecommunications services. It has published an informational RFC on existing practices in this area and is due to issue of the PINT protocol.

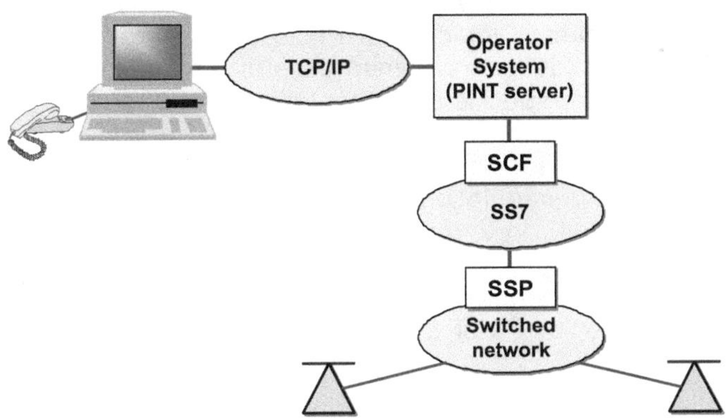

Figure 3.3 - The PINT architecture.

The PINT protocol enables the invocation of telephony services from terminals in an IP-based network environment (Figure 3.3). More specifically, a host in the IP network forwards a service request to a PINT server, which relays the request to the relevant PSTN network resource, such as a node implementing a service control function (SCF), which then executes the requested service, possibly reporting the service session status back to the originating IP terminal. The PINT protocol focuses on a small number of key services, namely request-to-call, request-to-fax, and request-to-hear-content. The protocol is specified as a usage profile of the session initiation protocol (SIP), specific extensions to SIP, and the associated standard session description protocol (SDP).

While the overall aim of the PINT initiative is to enable integration of Internet resources and telephony services in broad terms, it will effectively standardize access from the Internet to the IN service control function (SCF). It also enables development of novel services that are executed partially in the Internet domain and partially in the traditional telephony domain. Another consequence is that due to the reuse of standard IETF protocols and methodologies, the solution provided is very lightweight in nature. Importantly, the PINT protocol fits into the existing SIP architecture for Internet-based media session control, which will be significant in the future if SIP forms the basis for IP telephony. From a broader business perspective, important factors pointing toward the likely success of PINT are the fact that the IETF standardization process is proven in producing timely, flexible, scalable, and extensible protocols, and that nearly every major telecommunications equipment vendor and operator participates in the process. In addition, PINT is being actively considered by ITU-T SG11 for inclusion in the IN CS-4 functional architecture.

A potentially significant drawback of the PINT work is that no standard application programming interface (API) is defined as part of the IETF process. This either leads to an emergence of a de facto standard API or a multitude of vendor-specific implementations with resultant porting difficulty for code to run on different products. Indeed, as seen with past IETF standards, the lack of specification rigor may mean that it will take several years of vendor implementation experience before a high level of interoperability is achieved. In terms of interoperability it is also noted that interworking between the PINT gateway and the SCF is currently unspecified, although it is potentially a useful interoperability interface. However, it may be considered by ITU-T SG11.

Another factor to consider is that PINT competes, albeit indirectly, with other initiatives such as SIGTRAN (addressing transport of SS7 protocols over IP) within the IETF and external initiatives like ETSI TIPHON, which is also addressing integration of IP telephony and IN [Bre00, Cha00, Den03].

3.2.2 *PSTN/IN Requesting Internet Service (SPIRITS)*

PINT prevents IP hosts from participating directly in call control. Because such a capability would greatly enhance the possibilities for hybrid Internet/IN services, the IETF has started the PSTN/IN requesting an Internet service (SPIRITS) working group to address the use of Internet resources by SCFs during service execution.

The services in the SPIRITS working group addresses how services supported by IP network entities can be started from IN requests, as well as the protocol arrangements through which PSTN can request actions to be carried out in the IP network in response to events (IN triggers) occurring within the PSTN/IN. SPIRITS concerns architecture and protocols for secure transport of IN trigger information (requests for actions, as well as plain event notifications, including parameters) from PSTN/IN to the IP network, and optional responses from the IP network back to the PSTN/IN.

The SPIRITS architecture includes three potentially independent entities (Figure 3.4):

- SPIRITS client,
- SPIRITS server,
- PSTN/IN requesting system.

The SPIRITS client is the entity that requests notification or some actions to be performed in the IP network. The SPIRITS server is the entity that receives notifications or requests from the PSTN/IN and optionally returns responses back to the PSTN/IN, while initiating execution of the services requested in the IP domain. The SPIRITS server and PSTN/IN requesting system both reside in the IP domain, with the PSTN/IN entity on the boundary between the IP and PSTN/IN networks. The presence of three independent parties implies a requirement to support complex trust models. Accordingly, the security architecture must support limited trust between the parties.

The parameters passed in any SPIRITS Service request are limited to information available from PSTN/IN entities. An example of such a service is Internet call waiting: on an incoming telephone call, an IP node is notified of the call and can then carry out some actions. Definition of any information or data within the PSTN is the responsibility of the ITU-T and so is out of scope for SPIRITS.

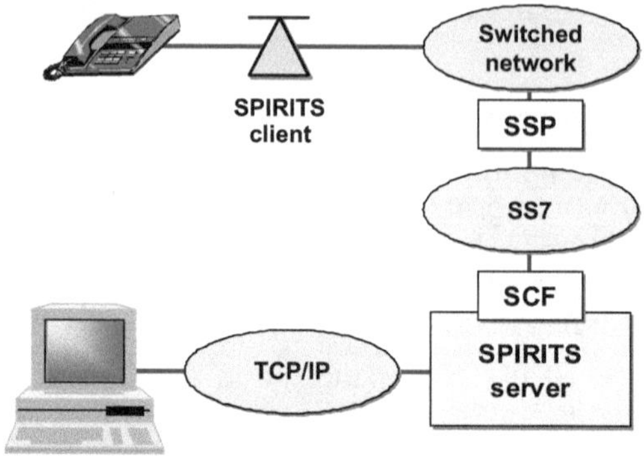

Figure 3.4 - The SPIRITS Architecture.

The target of this working group is to describe building blocks for PSTN-IP services that start from PSTN/IN requests, and not to standardize the PSTN-IP services themselves. The working group focuses on an event-oriented design, rather than a service-oriented design. Specific services to be considered initially are: incoming call notification (Internet call waiting); Internet caller-Id delivery; and Internet call forwarding and "follow me" [SPIRITS].

3.2.3 PARLAY

The Parlay industry working group was formed in April 1998 to specify an open API for telecommunications service control. A version 1 of the specification was issued in late 1998. The consortium is now working on version 2 of the Parlay specifications.

The Parlay group aims to specify an object-oriented service control API that is independent of underlying communications technologies (PSTN, wireless, and IP networks). The API is specified in the universal markup language (UML) and is designed to support all major middleware technologies (DCOM, CORBA, Java Platform). The technical goals of the group include encouraging computer-telephony integration (CTI), allowing IT systems of enterprises to access, control, and configure traditional telephony IN services, providing a unified abstract service control interface for heterogeneous media network types (PSTN, wireless, VoIP) and specification of value-added framework services such as online brokerage and billing mechanisms. Business goals include creating a market for third party service providers, enabling services that are more customized to individual enterprise needs, enabling smaller IT companies to develop telecommunications services and allowing network operators to sell access to their IN infrastructure.

The Parlay API enables a new generation of customer- or third-party-controlled services which are integrated into IT systems such as e-mail and customer information databases. These services will directly use the telecommunications IN capabilities operators without the need for wasteful call routing through private branch exchanges (PBXs) for redirection into the operator network (Figure 3.5). It is envisaged to promote rapid development of tailored services which allows use of general-purpose IT systems, thus reducing costs and increasing the availability of tailored services to small and medium-sized enterprises. The support of a large number of telecommunications equipment vendors, major network ope-rators, and, vitally, the dominant enterprise IT solutions provider indicates potential rapid proliferation of this technology. Careful consideration of exi-sting IN capabilities ensures that the API provides an easy evolution path from the traditional IN. Novel features of the API, such as online service brokering and framework interfaces for essential supporting functionality such as billing, combine to make a very complete solution. The API sup-ports the current business and regulatory drivers toward third-party ser-vice providers and the reselling of operator IN functionality.

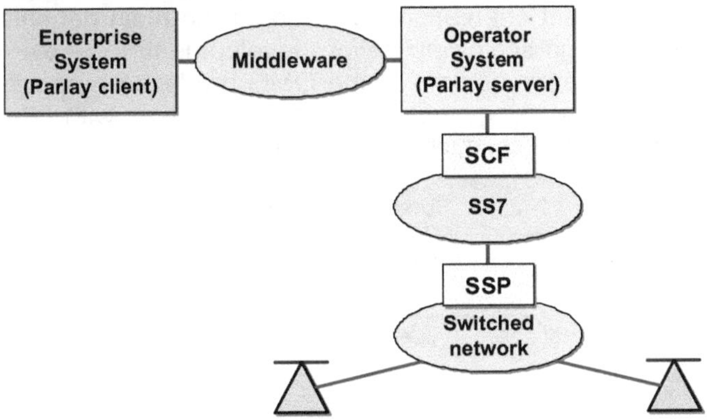

Figure 3.5 - The Parlay Architecture.

There are, of course, limitations to the work that the Parlay group has done so far. It is important to note that the current version 1.2 of the API includes large areas for further study. This includes many of the operational, administrative, and management (OA&M) features that make the solution so attractive. It is unlikely that there will be any clear winner of the current competition for dominance of the middleware marketplace, especially in the problematic domain of real-time service control. The technology-independent approach of the consortium guards against this, but will provide interoperability problems as multiple vendors claim Parlay compliance but only support one (or a subset) of the possible middleware implementation mechanisms.

Of course, the non-specific middleware approach also guarantees that while Parlay may be implemented with any technology, it is unlikely to be optimal since compromises must be made to ensure cross-technology support. It is apparent that the promise of an abstract service control interface for heterogeneous connections is not fully realized in the current specifications. There will be many challenges for future development in this area since very dissimilar call control mechanisms, such as SIP and the intelligent network application protocol (INAP), are integrated in one API. The result may well be that only very basic call control may be operated over such connections [Bre00, Cha00, Den03].

3.2.4 IN/CORBA Interworking

CORBA is software architecture, defined by the OMG, which enables software objects to interact with each other despite their location, type of host computer, or programming language. Improved system scalability, increased software reuse, ease of distribution, implementation language independence, and object orientation are seen as the main benefits of adopting CORBA for large-scale application development. In September 1998 the telecommunications domain of the OMG produced a specification focusing on the interworking of CORBA-based systems with telecommunications signaling systems, such as IN and mobile systems (Figure 3.6).

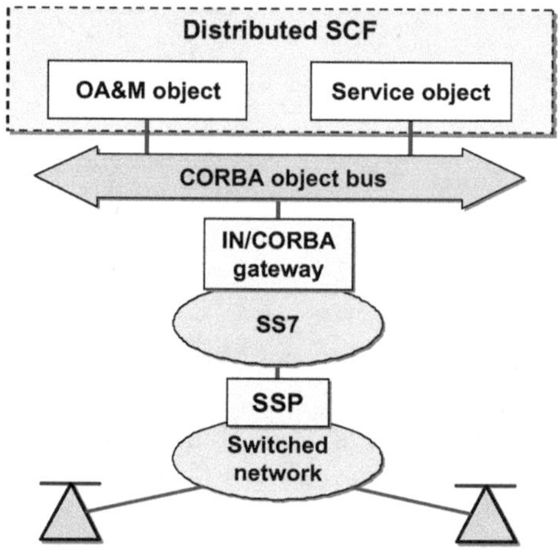

Figure 3.6 - IN/CORBA Interworking.

The primary technical motivation for the IN/CORBA interworking specification is to provide mechanisms for interworking of the existing service infrastructure, which uses transaction capabilities (TCs) for commu-nication, with CORBA-based service objects, which use an inter-object re-quest broker protocol (IOP) for communication. The specification defines a framework for the design of CORBA-based TC-user applications, such as INAP, which may communicate via a gateway with legacy TC-users such as service switching points (SSPs). An additional part of the speci-fication allows interworking between islands of CORBA-based systems using the existing signaling system 7 (SS7) infrastructure as a transport network for CORBA messages between, for example, switches that ex-pose CORBA interfaces and CORBA objects providing service logic.

Middleware technologies like CORBA are increasingly seen as the appropriate infrastructure for future service networks due to the inherent technological advantages brought to bear by distributed object-oriented processing environments. Indeed, CORBA has already been recognized by the American national standards institute (ANSI) T1 Committee as the basis for defining a framework and generic network information submitted to the ITU-T in an attempt to develop international standards for CORBA-based network management interfaces. In the same way, the CORBA interfaces provided by the IN/CORBA specification provide standardized interfaces which allow more open and distributed implementations of IN services. Also, the common CORBA approach to management and service provisioning produces a more integrated network and less cumbersome service management. IN/CORBA also facilitates increased interconnect-tion capabilities with external resources such as the Internet and private databases. The approach has the added advantage of providing a homogeneous interface for any SS7 protocol stack implementation. This independence reduces technology investments, allowing service creation to be independent of proprietary SS7 protocol stack implementations. OMG standards also have the desirable feature of fast standardization and short time to market. Because the specification enables implementation of both CORBA-based IN and MAP systems, it may also provide a common ground for fixed-mobile convergence. Leverage of existing off-the-shelf CORBA services, such as security, naming messaging, and notification, can also help accelerate application development.

Although the IN/CORBA approach presents many possibilities for the future of IN, there are some associated drawbacks. In order to maintain generality, the solution is quite low-level and does not provide support specifically for IN service development. CORBA still has shortcomings when expected to operate in a highly fault-tolerant real-time environment, as is expected of telecommunications systems. Indeed, these issues are not addressed directly by the IN/CORBA interworking specification. Since CORBA standardization is controlled by an IT industry body rather than a telecommunications body, it may be difficult to impose telecommunications systems requirements on standards based on an architecture for implementation of more general-purpose distributed systems [Bre00, Cha00, Den03].

3.2.5 *Telecommunications and Internet Protocol Harmonization Over Networks (TIPHON)*

The provision of voice-band services on technologies other than circuit switching has been growing over recent years with much effort on the provision of such services on packet switching infrastructures using the Internet protocol. TIPHON exists within ETSI to ensure that the requirements for quality of service, security, inter-domain settlement, and so forth, that arise from the abstraction of service from underlying technology applies equally to switched circuit network (SCN) and to packet switched technologies.

There is a growing market for real-time voice communication and related voice-band communication over Internet protocol (IP) based networks. The objective is to support a market that combines telecommunications and Internet technologies. The overall structure of TIPHON can be summarized in the Figure 3.7.

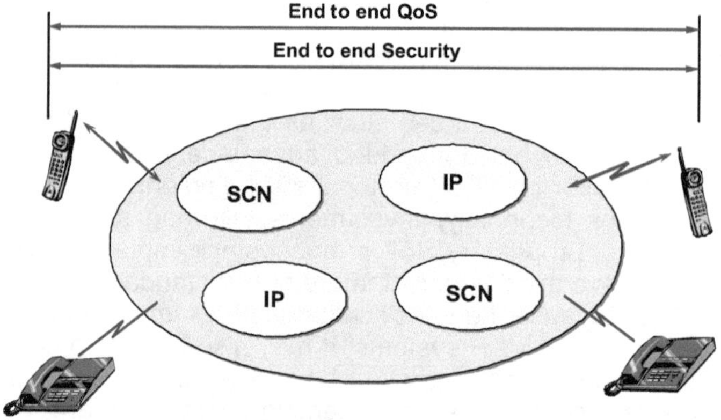

Figure 3.7 - Overview of TIPHON problem domain.

TIPHON can therefore be drawn as a network of networks where the constituent networks may be based upon IP or circuit switching technologies. The TIPHON service lies over these networks and establishes a means of providing guaranteed end-to-end quality of service (QoS) and consistent inter-domain security capabilities. In addition TIPHON ensures that service users and providers are able to call upon standardized inter-domain settlement protocols.

TIPHON services must be treated in like manner to existing regulated speech services and therefore encompasses provision of facilities that ensure compliance with national and regional regulations for privacy, including data protection, lawful interception, and accountability. In addition TIPHON is developed to meet the requirements arising from the provision of lifeline services which include availability and integrity services. Finally TIPHON also ensures that national and international regulations, current and planned where practicable, for subscriber number and service portability are supported.

The following assumptions are applied as guiding principles for TIPHON:

- TIPHON terminals may be PC-like or telephone-like;
- operation of a TIPHON terminal tends towards that of a telephone and therefore encompasses single stage dialing, network type abstraction;
- subscribers may move their access technology yet retain the same grade of service and same QoS.

TIPHON does not specify the use of any existing or new bearer services. However it is able to request certain QoS constraints that may restrict the ability of any particular bearer service to support TIPHON teleservices (particularly with respect to QoS and bandwidth).

The following teleservices in IP networks are considered:

- point to point speech,
- point to multipoint speech,
- point to point user defined data transfer, and
- point to multipoint user defined data transfer.

TIPHON restricts the network layers to be converged to (broadly) two options:

- Internet Protocol (IP), and
- Variants of digital signaling system number 1 (DSS1).

It is considered for simplicity that the various private networks technologies using PSS1 can be treated as variants of DSS1, which is itself the foundation of ISDN which in turn is the embracing technology for GSM, TETRA, DECT and ISDN itself [ETSI99].

3.2.6 ITU SG-11, SG-16 and H.323

3.2.6.1 Study Group - 11

Study group 11 provides technical advice concerning the allocation of signaling area/network codes. Close coordination with study groups of ITU-T and ITU-R as well as outside forums and consortia would continue to be maintained.

Defining signaling requirements and protocols for:

- telephone and N-ISDN basic and supplementary services,
- B-ISDN and multimedia services,
- user mobility and terminal mobility for UPT and mobile (FPLMTS) services,
- access and network security,
- control of transmission equipment (e.g., echo controllers).

Using technologies such as:

- intelligent network,
- signaling system No. 7, including MTP, ISUP, B-ISUP, SCCP and TC,
- digital subscriber signaling 1 (DSS 1) for N-ISDN,
- digital subscriber signaling 2 (DSS 2) for B-ISDN,
- data link layer for DSS 1 and DSS 2,
- ATM adaptation layer.

Supported by framework and methodology studies on:

- signaling and protocol framework for an evolving environment,
- unified functional specification methodology.

3.2.6.2 Study Group - 16

Responsible for studies relating to multimedia service definition and multimedia systems, including the associated terminals, modems, protocols and signal processing, it produces recommendations of H-series.

- Recommendation H.323 - This recommendation covers the technical requirements for multimedia communications systems in those situations where the underlying transport is a packet based network (PBN) which may not provide a guaranteed quality of service (QOS). These networks may include local area networks, enterprise area networks, metropolitan area networks, intra-networks, and inter-networks (including the Internet). They also include dialup connections or point-to-point connections over the PSTN or ISDN which use packet based transport such as PPP. These networks may consist of a single network segment, or they may have complex topologies which incorporate many network segments interconnected by other communications links.

The recommendation describes the components of an H.323 system. This includes terminals, gateways, gatekeepers, multipoint controllers, multipoint processors, and multipoint control units. Control messages and procedures within this recommendation define how these components communicate.

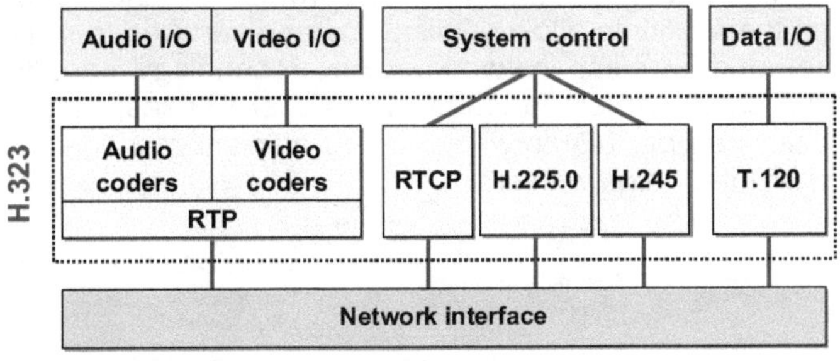

Figure 3.8 - H.323 architecture.

H.323 terminals provide audio and optionally video and data in point-to-point or multipoint conferences. Interworking with other H-series terminals, GSTN or ISDN voice terminals, or PSTN or ISDN data terminals is accomplished using gateways (Figure 3.8). Gatekeepers provide admission control and address translation. Multipoint controllers, multipoint processors and multipoint control units provide support for multipoint conferences. The scope of H.323 does not include the network interface, the physical network, or the transport protocol used on the network [ITU-H3].

3.2.7 JAIN

The objective of the JAIN initiative is to create an open value chain from third-party service providers, facility-based service providers, network providers, and network equipment manufacturers to carriers, consumers and computer equipment manufacturers.

The JAIN initiative integrates wireline, wireless, and packet based networks, as illustrated in Figure 3.9. The adaptation of network specific protocols to the JAIN model is covered in the protocol API specifications. Additionally, the JAIN initiative abstracts the protocols covered by the API specifications into a single call control, coordination, and transaction model to be used by compliant services.

The JAIN initiative brings service portability, convergence, and secure network access to telephony and Internet networks. This will positively alter the current business structure of these networks as follows:

- Service portability (write once, run anywhere) - technology development is currently constrained by proprietary interfaces. This increases development cost, time to market, and maintenance requirements. With the JAIN initiative, proprietary interfaces are reshaped to uniform Java interfaces delivering truly portable applications.
- Network convergence (integrated networks - any network) - by delivering the facility to allow applications and services to run on PSTN, packet and wireless networks, JAIN technology speeds network convergence. As demand for services over IP rises, new economies of scale are possible as well as more efficient management and increased integration with IT.

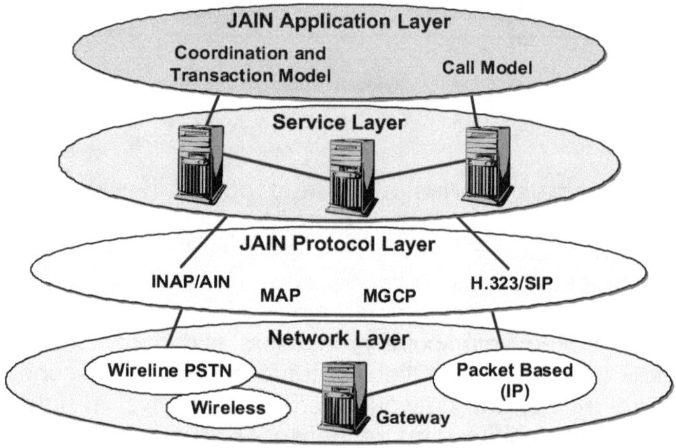

Figure 3.9 - JAIN initiative.

- Secure network access (by anyone) - by enabling applications residing outside the network to directly access network resources and devices to carry out specific actions or functions, a new environment is created for developers and users. The market opportunity for new services is huge when controlled access is provided to the available functionality and intelligence inside the telecommunications networks.

The JAIN initiative takes the telecommunications/Internet market from many proprietary closed systems to a single open environment able to host a large variety of services. Java and JAIN technologies allow carriers to extend and enrich their services. JAIN technology makes next generation application development faster, simpler, and less expensive through the use of Java technology.

The next generation architecture provided by JAIN technology creates an environment for deploying new services. This model is best served when all network levels participate - hardware companies, stack providers, network equipment providers, service providers, and carriers. In the heavily competitive telecom market, the carriers that embrace these next generation capabilities will succeed by leveraging their ability to create new services to differentiate themselves from less successful competitors [JAIN00, Bak00].

3.2.8 Computer-Telephony Integration

As computer telephony (CT) becomes an integral part of the entire communications network including the Internet. There are increasing challenges to making diverse communication products work together. The ECTF (enterprise computer telephony forum) is focused on solving the technical challenges of interoperability.

The ECTF is a non-profit organization composed of computer telephony suppliers, developers, systems integrators, and users from the USA, Europe, and Asia/Pacific. Together they discuss, develop, and test approaches to successful multi-layer interoperation within the PSTN, IP, and enterprise information system environments. Successful multi-layer interoperation enables application solutions that can exploit the full range of contemporary communications capabilities while lowering costs for both developers and users. The ECTF technical committee has a worldwide scope and addresses global technical needs for:

- convergence of computing and telephony,
- interoperability of defacto and de jure computer telephony standards,
- consistency of computer telephony interfaces,
- availability of scalable, networked, extensible computer telephony platforms and applications.

Its goals are as follows:

- provide architectural frameworks for interoperability,
- foster efficient and effective development of computer telephony products and services,

- facilitate industry acceptance of interoperability through common implementation agreements,
- promote industry cooperation and exchange.

The technical committee has a number of working groups (WGs) and task groups that underscore the areas of ECTF interest, such as:

- administrative services,
- application interoperability,
- architecture,
- call control interoperability,
- computer telephony services platform,
- hardware components interoperability.

3.2.8.1 Technical Background

The current computer telephony (CT) marketplace is complex and broad-based. It includes a wide variety of application-specific, predominantly proprietary devices like interactive voice response systems (IVR), voice mail systems, e-mail voice gateways, fax servers, switches (PBX and CO), automatic call distributors, and predictive dialers. The complexity is compounded by a user desire to integrate these devices with computing environments like host-based computer systems, client-server systems, and desktop computer systems. Services are enhanced and new ones added using emerging technologies like voice compression and expansion, text-to-speech, automatic speech recognition, facsimile, fax-to-text, data/fax/voice modems, desktop telephony interfaces, hearing im-paired devices and screen based phones. Enterprise telephony networks have become too complicated to understand, manage or perform as ex-pected and the duplication increases the cost of implementation and ser-vice. Figure 3.10 shows an example of an enterprise network.

Furthermore, applications and devices are sold and supported by a diverse group of vendors including switch vendors, interconnect vendors, system integrators, telecommunication companies, computer companies, application tool-kit vendors, and telephone server vendors.

This complexity has created huge interconnection and interoperability problems dramatically slowing the growth of the CT market. Software developers want to create integrated applications by combining features, services and technologies as needed regardless of supplier, vendor, technology, or industry source. Market forces have created a need for a set of agreements on the many interworking issues; agreements that free the customer to select any hardware, any platform and any application, put them together and build new services; agreements that allow users to enjoy straight forward implementations and avoid duplication of hardware, services and administration; agreements that encourage the market to grow.

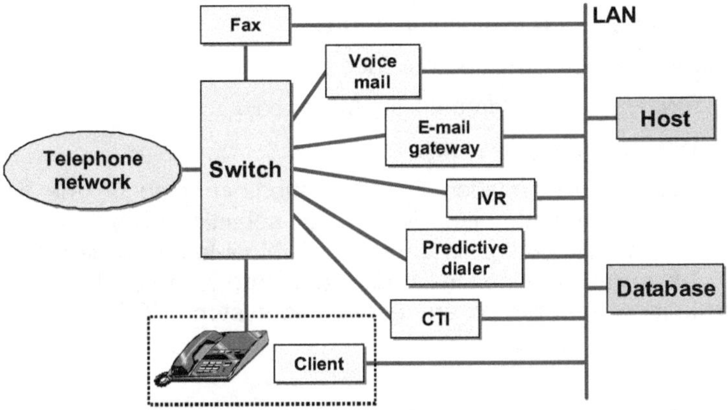

Figure 3.10 - Enterprise telephone network.

The task for ECTF is to make it easier for users to deploy CT services by understanding the myriad needs of users, service providers, integrators, distributors and suppliers. Then using that knowledge to identify, link and augment existing standards and publish implementation agreements to guide the CT industry toward interoperability. Interoperability is crucial to market growth since it offers lower cost products that are easier to install and maintain, are faster to market, and provide more options for custommers as well as suppliers.

3.2.8.2 ECTF Framework

The complexity of integrating CT subsystems can be reduced by defining general-purpose telephone components with fully specified interfaces to enable interoperability among different vendor's products. An open, modular framework that reduces the complexity of building and integrating CT applications will grow the market. Incompatible proprietary applications and devices will be replaced by more general and cost-effective solutions.

The ECTF framework defines two types of servers. Application servers execute telephone and media applications in a distributed network. CT servers provide the telephone and media resources (lines, voice recognition, fax) required by the applications.

By thoroughly specifying a few key interfaces between primary system components, the broadest range of interoperability can be achieved. The ECTF has defined the following interfaces:

- S.100: Media and Switching Services Interface,
- S.200: Transport Protocol Interface,
- S.300: Service Provider Interface,
- H.100: Hardware Compatibility Interface,
- M.100: Administrative Services Interface.

All the services are provided by applications running on the server or client systems. Implementation is cleaner, selection of modules is un-restricted, and administration is centralized. Modules from different suppliers can be mixed and matched, different hardware can be co-resident, and applications can run on any networked system [ECTF, Mur04, Bri05, Kry07a].

3.3 Call Models for Converged Networks

In Internet telephony, the call control functions of a traditional circuit switch are replaced by a device referred to, as call agent, SIP server, H.323 Gatekeeper, feature server, or softswitch. This device, which is re-ferred to as an Internet call agent, or simply a call agent, is an IP entity that coordinates the calls. A call agent executes a finite number of state transitions as it processes the call; these state transitions constitute its call model. The term call model when applied to an Internet call agent would be; better termed a protocol state machine. Unlike a traditional switch armed with an IN call model, the protocol state machine on a call agent does not contain IN specific triggers and states. Also, the number of call-related states of an Internet call agent are much less than those of the IN call model. Currently, there are at least two major Internet call signaling protocols in use - H.323 and SIP - both with varying number of states than the IN call model.

In order to access IN services transparently using Internet telephony, the Internet protocol state machine must be mapped to the IN call model. This has the added benefit of accessing existing IN services using the same detection points (DPs) from the same well known point in call (PIC). From the viewpoint of other IN elements like the service control point (SCP), the fact that the request originated from a call agent versus a call processing function on a traditional switch is immaterial. Thus, it is important that the call agent be able to provide features normally provided by the traditional switch, including operating as a SSP for IN features. The call agent should also maintain call state and trigger queries to IN-based services, just as traditional switches do.

The IN call model consists of two halves: the originating call model and the terminating call model. If the called and calling party share the same switch, the originating call model is assigned to the calling party and the terminating call model is assigned to the called party. If the call has to go through multiple switches to get to the destination, each of the intervening switch will run the two halves of the call model, with the destination switch's terminating call model providing services to the called party. While this model has worked well for traditional circuit-based switching, it may not be desirable to implement it in an analogous manner on an Internet call model.

The most expeditious manner for providing existing IN services in the IP telephony domain would be to use the deployed IN infrastructure as much as possible and leverage existing services. The logical point in the IP telephony domain to tap into for accessing existing IN services is the call agent. However, the call agent does not run an IN call model. Instead, the various Internet call agents run their respective native protocol state machines for call signaling - either Q.931 in H.323 or a SIP stack in SIP. The trick, then, is to overlay this state machine with an IN layer such that call acceptance and routing is performed by the native state machine and services are accessed through the IN layer using an IN call model.

3.3.1 Call Models Description

In a traditional switch environment, when the service switching point (SSP) recognizes a call that requires IN treatment, it temporarily suspends the call processing and sends a query to the service control point (SCP). The SCP analyzes the information it received from the SSP and makes a decision on how to continue processing the call. The decision is sent to the SSP, which now continues with further call processing. It is important to realize that IN treatment for a call is not limited to simple request-reply transactions. Including simple querying, the following are the major functions that are part of ITU-T capability set 1 and 2 (CS-1 and CS-2): querying, caller interaction, trigger activation/deactivation, response processing. In converged networks, there are two call models, which have to interwork for hybrid services support:

- IN call model - The IN generic basic call state model (BCSM), independent of any capability sets, is divided into two halves - an originnating call model (O_BCSM) and a terminating call model (T_BCSM). There are a total of nineteen PICs and thirty-five DPs between both the halves (eleven PICs and twenty-one DPs for O_BCSM; eight PICs and fourteen DPs for T_BCSM). The SSPs,

SCPs, and other IN elements track a call's progress in terms of the basic call model. The basic call model provides a common context for communication about a call.

- SIP call model - SIP is a lightweight signaling protocol for Internet telephony. SIP has six messages (INVITE, ACK, OPTIONS, BYE, CANCEL, and REGISTER) and various response codes belonging to the following six classes:

Class	Meaning
1xx	Informational
2xx	Success
3xx	Redirection
4xx	Request fail-ure
5xx	Server fail-ure
6xx	Global failure

Table 3.1 - SIP response codes.

For mapping the IN call states, the SIP protocol state machine can be viewed as essentially consisting of an INVITE message, interim response codes for the invitation (100 Trying or 180 Ringing), an acceptance (or a decline) of the INVITE message, and an acknowledgement for the acceptance (or decline). If the invitation was accepted, SIP provides a BYE message for signaling the end of the call [Wan07].

3.3.2 IN Call Model Mappings

One way in which IN services can be invoked transparently from a SIP server processing a telephone call is to overlay the SIP protocol state machine with the IN call model. Thus, the call receives treatment from two call models, both working in synchrony; the SIP state machine handles the acceptance and final delivery of the call, while the IN call model interfaces with the IN to provide services for the call. Figure 3.11 demonstrates this concept:

- SIP server accepts a call and notifies the IN call handling layer of this event;
- IN call handling layer interfaces with the IN elements to provide services for the call, ultimately informing the SIP server on how to deliver the call.

The underlying assumption is that IN is servicing the call by providing it features, and SIP is simply routing the call based on the decisions made by the IN layer.

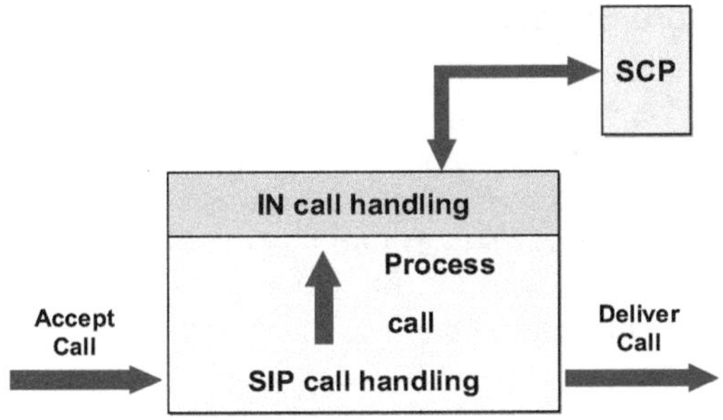

Figure 3.11 - IN call model overlayed on SIP.

Another fundamental problem lies in the notion of a call state. The IN call model is necessarily a stateful one. A SIP server can operate in either stateful or stateless mode, depending on the needs of the application. For speed, reliability and scalability, SIP servers may be run in the stateless mode. The duration and amount of state maintained at a SIP server are small compared to the traditional telephone network, where the switching node must maintain the call state for the entire duration of the call. For a SIP server to run in the call-stateful mode, it has to indicate a willingness to remain in the signaling path till the call is disconnected. This is accomplished using the record-route header field of a SIP message.

3.3.3 IN/SIP Architecture

In order to apply the stateful IN call model to a SIP server, the originating and terminating SIP network servers must run in a call-state aware mode and have the IN call model layer working in conjunction with SIP as depicted in Figure 3.11. Other intervening SIP servers may remain stateless and have no need to run the IN call model layer. The originating and terminating SIP network servers simulate the originating and terminating switching nodes in a traditional telephone network. IN services accessed through detection points (DPs) on originating or terminating side can now be handled by the IN layer on the originating or terminating SIP server. Figure 3.12 demonstrates this. There are three other points to be mentioned in Figure 3.12:

- If the called party and the calling party are handled by the same SIP server, both halves of the IN call model will run on that server. This is analogous to the traditional telephone network.
- In the traditional telephone network, the inter-exchange switch nodes can run both halves of the call model. This can also be accomplished in the SIP network if desired. Figure 3.12 shows the IN call model running on originating and terminating SIP servers. However,

any of the core SIP servers could also have hosted the IN call model if needed.

- If the called party and the calling party are handled by different SIP network servers, each with its own IN layer, the IN call state information has to be communicated between these servers. Or in fact, the IN layers of the originating and terminating SIP servers can communicate directly with each other using ISUP over IP to share call state between themselves.

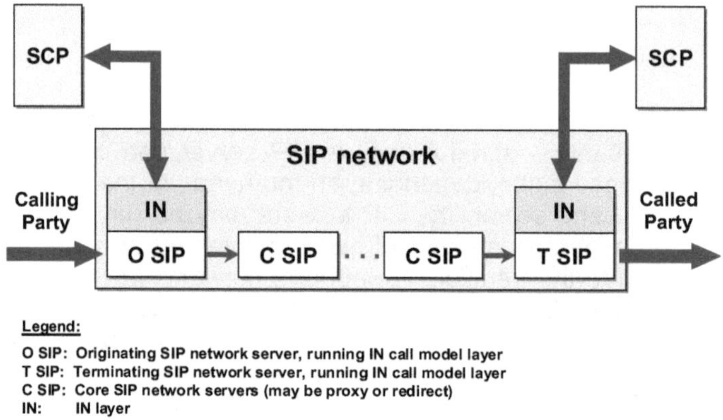

Figure 3.12 - IN-controlled SIP network.

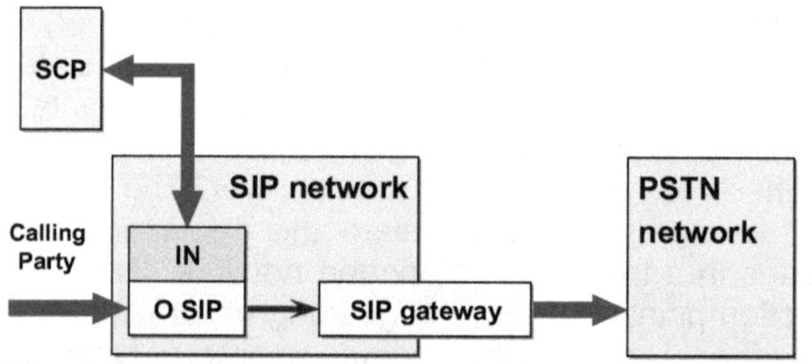

Figure 3.13 - SIP network with PSTN gateway.

Figure 3.12 shows an end-to-end SIP network, with SIP servers running the IN call model reaching out to the SCP for service logic. Figure 3.13 shows a SIP network providing services from the SCP through the IN call model and routing the call to the telephone network. In this example, it is assumed that both halves of the IN call model are running on the same SIP server [Gur00, Fat06, Kry06b, Kry08a].

3.4 Service Control for Voice and Data Networks

The service control is depicted in Figure 3.14. It is a collection of functionalities that interact due to the communication transport services as provided by a distributed processing environment (DPE).

The composition and interactions of functionality allow service designers to build added-value telecommunication services without the need of a deep knowledge of underlying network and special resources. In addition, this layer may provide service features to external domain services/ applications. These features are offered to third party service providers by means of secured/controlled interfaces in order to ensure secure, robust, accountable interactions. Basic functions that must be supported by this layer are: call control, messaging control and service session.

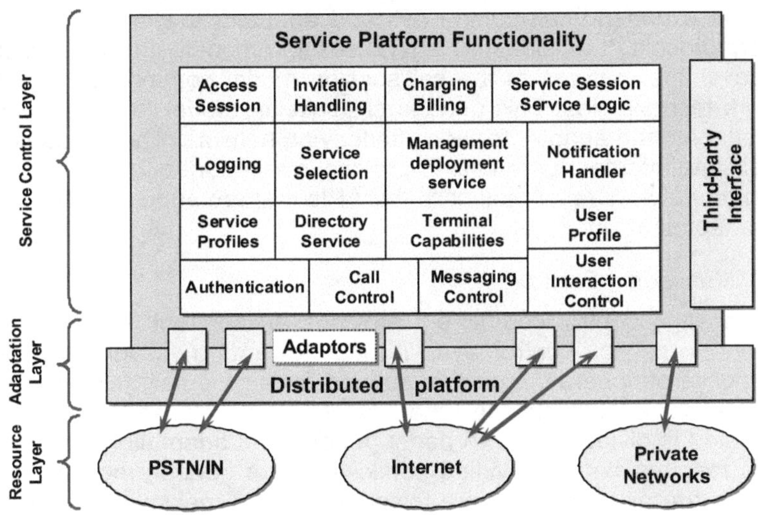

Figure 3.14 - Framework for the service platform.

The network and special resources can activate services implemented in this layer by means of events that are sent to the service platform. Resource adaptors or (resource) wrappers are used to hide the details of specific protocols (e.g., INAP, Megaco/H.248) from the upper layer.

Software components are to be defined to implement the supported functionality. For the component model it could be useful to adopt a componentware model (e.g., a JavaBeans-like model): in this case for each component it is important to identify the offered interfaces and the emitted notifications/events. The definition of the software components according to a componentware model enforces a greater degree of independence among the components implementing the basic functionality.

3.4.1 Call Control

The call control function allows to control calls in heterogeneous networks. The call control offers a common call control interface to all services. The call control component is responsible for the mapping between the set of operations offered to the services and the protocol specific operation (or sequence of operations) supported by the underlying networks adapters. Call control also has the role of handling hybrid call legs (call legs connected to different transport technologies/ network domains, i.e., like in PSTN-to-VoIP calls).

The call control functionality has two parts. A generic part provides services via a generic call control API and adapters towards different networks/protocols. It includes a call and session model, simplified so that it can cover the features of the call/session model common to underlying network technologies. This generic call/session model API is mapped, if needed, onto the adapter for each underlying network. These adapters also translate internal operations to protocol dependent operations need-ed to control network resources or to the APIs that are supported by the network adaptation layer.

3.4.2 Messaging Control

The message control component provides an abstract interface to services in the service control layer. It maps towards messaging equipment and special peripherals providing support for e.g., email, fax, text-to-speech, speech recognition and other features. The messaging control like call control is divided into two parts: generic and adaptation. The generic part is responsible of providing services with a generic messaging API. Adapters are also needed here towards different messaging systems involved. Messaging systems could comprise servers for e-mail, voice-mail, fax, and SMS.

Other functionalities of the service platform are:

- Directory service - supports information retrieval on users (e.g. user related data, services);
- Charging and billing - permits the provider to charge users/providers for service/functionality usage;
- Authentication - provides support for user authentication towards the service platform provider;

- Authorization - allows users (or third party providers) to have access to other functionality supported by the service platform upon successful authentication;
- Logging (mainly for internal use of the platform provider) - supports other functionality such as authentication, charging, billing, and auditing;
- Management (service deployment) - supports service and user's management from the platform provider;
- Web server - allows navigation, retrieval of information that can be either locally or remotely stored (proxy);
- Invitation handling - represents the capability for a user/third party provider of sending and receiving invitations;
- Service selection - allows a user/third party provider to select the service to be used, providing its authorization;
- Service logic/service session - offers capabilities to execute services/service features (e.g., handling of multiple instances of a service, service instance context, etc.) and to co-ordinate the usage of supporting functions;
- User profile - gathers user personal data (name, address), user's logical identifiers (logical name, personal number, e-mail address), authentication data (login/password, personal number/PIN), user's preferences (preferred language, billing info, etc.);
- Notification handler - provides a supporting component that co-ordinates the dispatching of events/communication exchanges among the other functionality;
- Terminal capabilities - describes available characteristics of terminals;
- Additional functionality - comprises customer service management (including personalization, registration), virtual presence call control (dispatcher), etc. [EUR909, Lul06].

3.5 JAIN APIs for Converged Networks

Java is designed to operate in distributed environments, which means that security is of paramount importance. With security features designed into the language and run-time system, Java allows to construct applications that can not be invaded from outside. In a networked environment, applications written in Java are secured from intrusion by unauthorized code. Used along with a firewall, Java is the basis for securely opening IN to data networking.

3.5.1 JAIN SS7 APIs

Substantial investment has been made in writing software for the SS7 protocol layers in the current network infrastructure. Most of this software is written in native programming languages, such as C, which prevents fast and easy introduction of new services in IN and wireless networks. This is because programs written in these languages cannot be easily uploaded to a running system. To upgrade the service logic in a service control point (SCP), usually one must take down the SCP, install the new software, and then bring it back online. Java, as an object-oriented programming language inherently supports distributed, platform-independent computing, this providing the framework to easily upload new objects containing various service components into the existing infrastructure. In order to utilize Java technology for this purpose, new service applications written in Java must be adapted to communicate with protocol layers that are implemented in programming languages such as C. The JAIN SS7 API is an effort toward providing such an adaptation tool.

Due to various market forces and regional requirements, present telecommunications networks embody many variants for each protocol layer. Whenever a new protocol variant is introduced into the network to provide a new bearer service or some other advanced network-level service, application-level software entities must be upgraded to work with the new protocol variants. The JAIN SS7 API is an attempt to abstract the network-level functionality from the peculiarities of different protocol variants so that changes in the underlying networks are transparent to the application.

At its core, the JAIN SS7 API architecture defines a set of software components that enable an application (e.g., service execution environment in an SCP and call control in an SSP) executing in the Java space to access services provided by the underlying SS7 protocol layers written in another programming language. IN service components, such as service independent building blocks (SIBs), are analogous to objects, or JavaBeans. The service creation and management center will create and upgrade the SIBs in Java and upload them at the service execution environment in real time using either the Java dynamic management kit (JDMK) or the enterprise JavaBeans environment. In addition, the JAIN SS7 API provides an operation, administration, and maintenance (OA&M) API to administer and manage various SS7 protocol layers.

The software components defined in the JAIN SS7 API are based on the JavaBeans design pattern. Each JAIN SS7 API defines three major Java software components stack, provider, and listener. These three software components are defined in the form of Java interfaces. The stack component abstracts the underlying native SS7 protocol stack, and provides a factory to create and manage the provider component, which exchanges protocol messages with the original SS7 protocol layer using a proprietary interprocess communication (IPC) mechanism. Therefore, provider components are specific to a particular implementation of a protocol layer. Listener components interact with the provider component via event ob-jects. To exchange event objects, listener components must register with provider components. Listener components should be portable across various implementations of provider components. Applications use the JAIN SS7 API to implement a listener interface. The provider component maps the generic API to the original protocol stack.

The JAIN SS7 API defines Java-based APIs for the following protocols:

- TCAP - a non-call-related transaction-based control protocol that provides support for message exchange between interacting applications in a distributed environment, such as the SS7 network infrastructure.
- ISUP - a call-related signaling protocol that provides support for call establishment and release and trunk circuit management in an SS7 network.
- MAP - a non-call-related control protocol that provides support for interacting mobile applications in a distributed network environment.
- INAP - a non-call-related control protocol that allows applications to communicate between various functional entities of an IN. The protocol defines the operations required to be performed between service providers for providing IN services, such as number translation, time of day, and follow me. The JAIN INAP API will be based on the American national standards institute (ANSI)/Bellcore advanced intelligent network (AIN 0.2) and the CS-2 INAP specifications of the international telecommunication union.
- OMAP - a standard mechanism used to provision and maintain various protocol entities in the SS7 network.

3.5.2 JAIN IP APIs

To provide services over an IP network equivalent to those of a traditional telephone network, various new protocols, such as H.323, media gateway control protocol (MGCP), session initiation protocol (SIP), and various new network entities, such as media gateways, gatekeepers, call agents, and media gateway controllers, have been defined.

There are a wide variety of devices and platforms that can be used to provide multimedia services over IP networks, depending on the combination of functions deployed in a single device. However, the services that are required by end users are well known; for example, for telephony, services such as call waiting, caller identification, 800-number translation, and calling cards are required. It is desirable to allow the services to be written in Java so that they can migrate across a variety of platforms as IP telephony networks and technologies evolve. By defining Java APIs to use traditional protocol software, the JAIN IP API subgroup seeks to preserve the existing protocol software in these network entities, while simultaneously enabling new Java applications to provide new services. The JAIN IP API will define Java-based APIs for the following protocols: H.323, MGCP, and SIP.

The generic Java-based interface that sends and receives SIP control messages to and from the native SIP protocol stacks in SIP clients and SIP servers. The JAIN SIP API are based on RFC 2543.

3.5.3 JAIN APIs for IN/IP Convergence

JAIN is suited to enable applications to migrate smoothly across diverse networks. As the JAIN APIs abstract the network details, the applications become network-independent. The following examples illustrate the applicability of JAIN in this context.

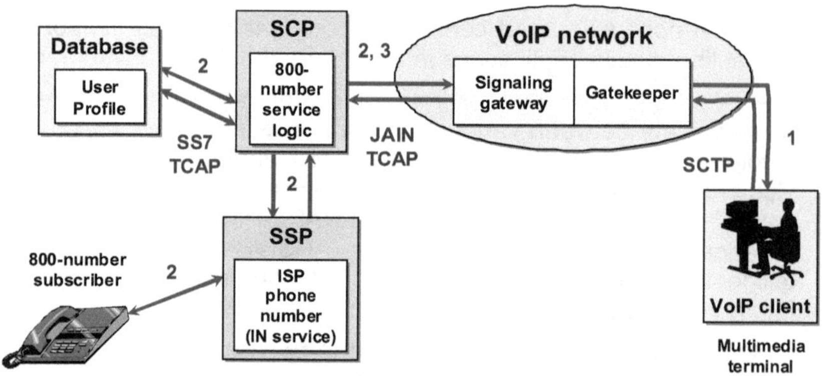

Figure 3.15 - ICW – Connection to Internet.

3.5.3.1Access of IN services from voice over IP network

An example relates to access of IN services, such as 800-number translation, from voice-over-IP (VoIP) networks (Figure 3.15).

- Step 1: The multimedia terminal (VoIP client) initiates a call to an 800-number in IN. The VoIP call request is typically routed by a gatekeeper element in the VoIP network.
- Step 2: This will then need to access data in the SCPs. Such access is typically achieved using the SS7 TCAP protocol. However, a gatekeeper application written in Java could use a JAIN TCAP API to

communicate with the database. The TCAP protocol can run over traditional SS7 transport, requiring the gatekeeper to have an SS7 interface.

- Step 3: This will imply that the gatekeeper and the signaling gateway are collocated. Alternately, TCAP can run over an IP network, using SCTP, to a signaling gateway, which then provides access to the database. Thus, in this scenario, the gatekeeper application can provide SS7-based services while running on top of an IP network.

3.5.3.2IP as a transport network for IN services

This example involves an IP as a transport network for IN services (Figure 3.16).

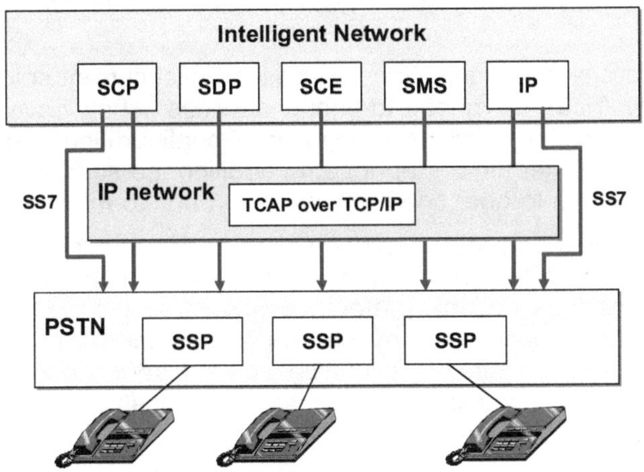

Figure 3.16 - ICW – Connection to Internet.

- In this case, the traditional IN elements such as SCP, SDP, service creation environment (SCE), service management system (SMS), and IP are not connected over traditional SS7 transport; rather, they are connected to an IP network.
- TCAP runs over TCP/IP and serves to carry the IN messages over the IP network.
- The SCP and IP still have an SS7 interface in order to connect to the PSTN equipment.
- However, IN service creation, administration, and execution can be done over the IP network at the back end. For example, the SCP with the SS7 interface could have the JAIN TCAP API implemented.

- The service logic that uses the API can execute in a remote server using an abstraction such as Java remote method invocation (RMI) or common object request broker architecture (CORBA) that runs over the IP network. This allows the service modules to be distributed across the network, instead of inside a single device, and the modules can be upgraded dynamically without affecting the operation of other service modules.

As network technologies and protocols proliferate, the ability to write Java-based applications that are portable across various networks will become increasingly attractive and cost-effective [Bha00, Kry08a].

3.6 Services that can be Recognized as Converged

The services supported by converged networks fulfill the following requirements:

- Any-to-any communication services - Whilst current solutions for IN mainly focus on phone-to-phone services. They have to support communication services in any kind of configuration (e.g., phone-to-PC, "any IP terminal"-to-phone. In addition the supported topologies must be one-to-one, one-to-many, and many-to-many.
- Customer-centered services - A user is represented in the network by a user profile that refers to all the relevant information concerning personal subscriptions, preferences, constrains, etc. The user profile has to be shared by any service. Furthermore, additional preferences and constraints could be derived from the composition of profiles into groups and business relationships (e.g., user/subscriber).
- Mixed voice/data services - Services could combine voice-related features and data-related features. The services could enable communication between persons, machines, and applications. The quality of service has to match and even bypass the current one. QoS should be negotiable in order to match the requirements of users for any single instance of communication.
- Seamless service access - A user must be able to access to the subscribed services by means of any terminal (fixed phone, mobile phone, PC, PDA, etc. according to the terminal capabilities), in the same way, and with the same subscribed features.
- Application-network synergy - Services could be activated by applications (e.g., Internet/Telecom services, PINT-like services, SPIRITS-like services) and could interact with application/systems,

possibly deployed in third party administrative domains (e.g., enterprises, added value service providers).

In the next subsections, service descriptions that can be recognized as converged are presented [Min00, Kry07a].

3.6.1 Overall Services Description

The overall services description is as follows:

- Internet Call Waiting - Internet call waiting service enables a user engaged in a dial up Internet session to be alerted when an incoming call has arrived. After the Internet user has been alerted, he is given several options for handling the call e.g., forwarding it, sending a waiting announce/tone, accepting the call over telephone network suspending the Internet session, or accepting the call over IP keeping alive the Internet session.
- Virtual Second Line - This service allows the subscriber to answer incoming phone calls while his single telephone line is busy due to an ongoing Internet session. A vocal gateway can be used to transform the incoming telephone call into a voice over IP flow direc-ted to the terminal interconnected to the Internet. In this way, the terminal could manage the IP flow carrying the voice along with the other IP flows originated by the web surfing.
- Click-to-Dial (Request-to-Call) - A user is able to initiate a telephone call by clicking a button during a web session. The called address (as well as the caller address) is either an IP address or a phone line number. The charging party could be either the initiator or one of the called parties.
- Distributed and Enhanced Call Center - The integration of IN and Internet would allow to decrease the costs of the call centers. A first opportunity is the implementation of network-based call center solutions in order to achieve distributed/virtual call centers. With this approach, the actual necessary infrastructure in each call center could disappear, and some new paradigms like teleworking could be used.
- Meeting Scheduler - A service that enables a user to have a web based user interface to schedule a meeting. He decides the time and the attendants. The timing information is used by an external application that launches the service via the access function. Every meeting attendant is confirmed, either with an e-mail notification and a following confirmation process via a web page or with a phone call.
- Unified Communication - This service allows users to send, retrieve and receive messages disregarding the format and the terminal

where the user is connected. The user must be able to create and respond to multimedia messages from any terminal and create and send any type of message without regard to the recipient's mailbox requirements. The user must also be able to reply to messages and forward messages with calls, and reply and forward calls with messages.

- Virtual Presence - A service that allows its subscribers to be reached anywhere, anyway by using both asynchronous messages and real time communication and from any terminal independently of the type of terminal he is logged to. Virtual presence extends the characteristics of a personal number service to the context of Internet/ telecom convergence. The aim of the service is to provide an integrated set of features that enable for example a subscriber to control the incoming calls according to a set of personal screening/routing rules (a sort of net-secretary) and terminal registrations that he/she could dynamically modify.
- Virtual Private Network - The enhanced VPN service would allow using most of cases of the VoIP paradigm. The main characteristics of this service could be:
    - Integration of several terminals into the VPN: fixed and mobile phones, PC,
    - Several communication modes (phone-to-phone, phone-to-PC, PC-to-phone, PC-to-PC),
    - Use of traditional features in VPN service: unified billing, abbreviated dialing, call restrictions (extra/intra VPN calls),
    - The service logic could manage information about the connections between the VPN positions, in order to select the cheapest way to reach the destination of a call, either in an intra VPN call or in an extra VPN call [Bas08].

3.6.2 IN Service Requests towards IP Networks

Some IN services could be extended to notify events or send information to the Internet domain. It should enable services like Internet call waiting, virtual presence, etc. It must be possible to have access to the Internet domain from IN for the purpose of:

- invoking Internet services, e.g., forwarding of fax/SMS as email,
- notify call information,
- controlling VoIP resources,

- access to customer and service information stored on Internet databases/web server, e.g., downloading share information from a web server for transfer to a GSM phone using SMS.

The following services can well help in defining requirements in the context of how services supported by IP network entities can be started from IN requests.

3.6.2.1 Virtual Presence Service

The VP service may work with a heterogeneous set of terminals: e.g., fixed and mobile phones, fax and PC, and new generation terminals (PDA, WebPhones, WAP-based mobile phones). The association subscriber-terminal can be dynamically changed (personal mobility). In addition a subscriber can be registered on multiple terminals of the same type. A service subscriber has a single virtual identity in the system, referenced by a logical name, personal alias, (used from the Internet context) and a personal number (used both from IN context and Internet one).

More precisely the user can:

- register/de-register - he can choose the terminals where he decided to be contacted,
- configure - he can define the group of terminals of common use,
- set personalization rules - using a graphical application, he can introduce and manage filtering rules to be applied to incoming asynchronous or synchronous communication requests.

Rules are modeled according to the ECA paradigm, i.e., event-conditions-actions:

- An event is an external stimulus addressed to the subscriber. Conditions represent constraints on rules scope,
- Actions are operations performed if and only if all conditions of the rule are satisfied when the event occurs. Rules are activated by a single event; they are evaluated against a set of conditions, and trigger a sequence of actions.

Service features that have an impact on the PSTN-to-IP context are:

- Notifying the subscriber about incoming synchronous communication request on his/her PC - As an example considering user A who had previously set some customized rules for handling incoming communication requests from other users. A user B tries to contact user A from a phone by using his personal number. The incoming call request is processed by the service. The service analyses the rules and one of them fires (event and condition matches). For example the action associated to the rule is to route the call to the user A office PC. Then the service forwards the call to the PC of the subscriber who is requested to accept the call through some stan-dard API or network protocols (e.g., IETF SIP). After he clicks on the appropriate button the call is completed and the two users are put through.

- Notifying the subscriber about incoming asynchronous communication request (e.g., vocal message) on his/her PC - A user B tries to contact user A from a phone by using his personal number. The incoming call request is processed by the service. The service analyses the rules and one of them fires (event and condition matches). For example the action associated to the rule is to route the call to the home phone of user A but this is busy. The service then evaluates the rule associated to the busy condition and for example it forwards the call to voice-mailbox of the subscriber. He is then notified about this voice message [Jha04].

3.6.2.2 Internet Call Waiting

Internet call waiting (ICW) is a service that enables a user engaged in a dial-up Internet session to be alerted when an incoming call has arrived. After the Internet user has been alerted, he is given several options for handling the call (e.g., forwarding it, sending a waiting announce/tone, accepting the call over telephone network suspending the Internet ses-sion, or accepting the call over IP keeping alive the Internet session). In parallel the caller is announced that the callee is busy and (s)he is asked to hold the line. In order to enable the service the subscriber has to use a client software that performs the registration phase by storing the asso-ciation between the telephone network line number and the IP address of his/her Internet session.

Service features that have an impact on the PSTN-to-IP context are:

- Handling an incoming call - The subscriber receives a telephone call directed to his/her phone, which is busy, since it is engaged in a dial-up Internet session. The IN service switching point (SSP) tri-ggers the service logic to handle this call event. The caller is con-nected to a intelligent peripheral in order to send a message to inform him/her to wait (call queuing). The service retrieves the IP ad-dress of the callee and then, depending on the user profile, notifies the incoming call to the Internet user with caller number or name or other call relater information. The Internet user may choose different options:
    - Accept call on PC using voice over IP,
    - Accept call on phone,

- Suspend IP session and answer the call on the phone,
- Reply the caller with a pre-registered message,
- Reject the call.

3.6.2.3 ICW service – connection to the Internet

The ICW subscriber connects to Internet by means of a dial-up connection (Figure 3.17).

Pre-conditions: User A has subscribed to ICW service.

Post-conditions: Trigger detection point (DP) 13 (in ETSI core INAP terminology DP13 corresponds to TCalledPartyBusy event) is armed so that when there is an incoming call for user A and his line is busy the ICW is notified.

- Step 1 - The dial-up connection is established.
- Step 2 - User A launches the Internet connection software, which dials the ISP phone number, that is a number which represents an IN service. As a result of the pre-arranged agreement between the Internet service provider and the network provider the DP3 (Analyzed-Information) was set.
- Step 3 - The SSP triggers the service logic by sending an InitialDP message to the IN service control point (SCP). Then the SSP, through a gateway, notifies the ICW service logic about a network event related to an incoming call for the ISP phone number. The service logic is executed within a kind of SLEE (service logic execution environment) that provide API to interact with network functionality. The ICW service logic consequently sets TDP13 using some management interface on the SSP. The ICW service logic then subscribes to call event related to user A call link disconnection since it needs to disarm the TDP13, when the Internet connection is disconnected.
- Step 4 - The ICW service logic gives instruction to the network to route the call to the address of the network access server (NAS) to be connected. Furthermore it arms DP7 (OAnswer) in order to be notified about the result of the call routing.
- Step 5 - The connection is set up between the SSP and the NAS and consequently the network acknowledges the establishment of the connection. Since the ICW service logic needs to monitor NAS disconnection as well as user A disconnection, it subscribed to call event (DP9ODisconnect) related to NAS disconnection (this subscription could not be requested before the successful notification of

the call routing to the NAS). When the network acknowledges the successful establishment of the call link towards the NAS, the call processing is stopped at the DP7 (OAnswer) since it has been set as an EDP-R. The ICW service logic instructs the SCP to continue the call processing (this is mapped onto the INAP operation Continue).

- Step 6 - The subscriber is now requested to authenticate through a login and password. This subscriber data are sent to a RADIUS server to authenticate him/her and to retrieve from a directory server user's A profile based on its home phone number. The association between the IP address of user A and home phone number of user A is stored in his user profile. The software in charge of handling invitations is launched automatically after the connection to Internet is established e.g., IETF SIP UAC.

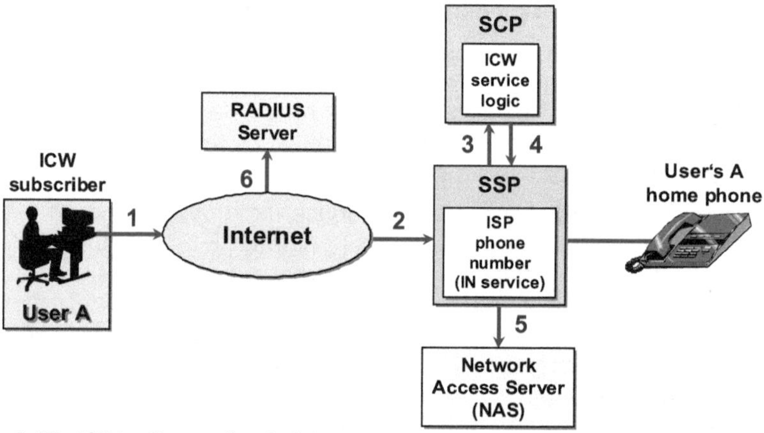

Figure 3.17 - ICW – Connection to Internet.

3.6.2.4 ICW – Incoming call notification

The user B tries to call user A at his home phone number (Figure 3.18).

Pre-conditions: User A is connected to Internet.

Post-conditions: User A is notified of an incoming call and has to choose either to reject the call, to answer the call over a PSTN line or to answer the call over IP.

- Step 1 - The user B dial user's A phone number.
- Step 2 - User's A SSP detects that user's A phone line is busy and according to the previous arming of TDP13.

- Step 3 - An InitialDP message is sent to the SCP in order to notify that a call has been triggered.
- Step 4 - An answer message is send from SCP to SSP.
- Step 5 - The call is routed to an interactive voice response (IVR), which implements IN special resource functions. The ICW service logic instructs to send a ConnectToResource message to the SSP. The SSP then establish the connection to the IVR.

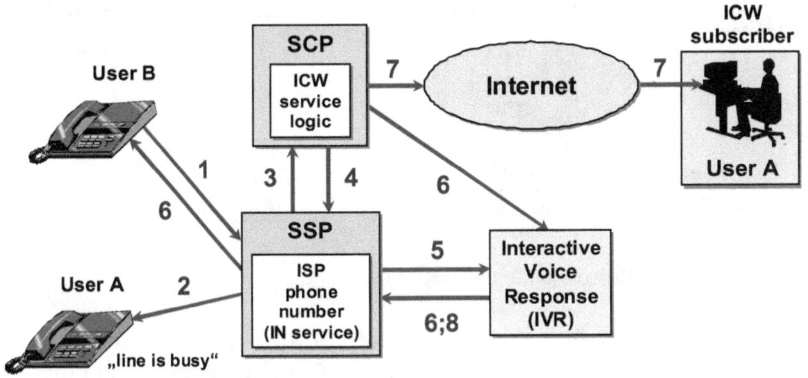

Figure 3.18 - ICW – Incoming call notification.

- Step 6 - The service logic sends the PlayAnnouncement operation to the IVR thus an announcement is played to the user.
- Step 7 - In the meantime (while the announcement is played or even while the connection to the IVR is established), the ICW service logic retrieves user's A profile from the directory server in order to get user's A PC IP address. The ICW service logic sends an invitation (e.g. SIP INVITE message) to user A with information about the caller (e.g. calling line identity) asking him either to:
    - Accept call on PC,
    - Accept call on phone,
    - Suspend IP session and answer the call on the phone,
    - Reply the caller with a pre-registered message,
    - Reject the call user's A choice is sent back to the ICW service logic via a SIP response message.
- Step 8 - The IVR reports the end of the announcement to the SSP.

3.6.2.5 ICW - Rejection the call

The user A is invited to a call by a user B and chooses to reject it (Figure 3.19).

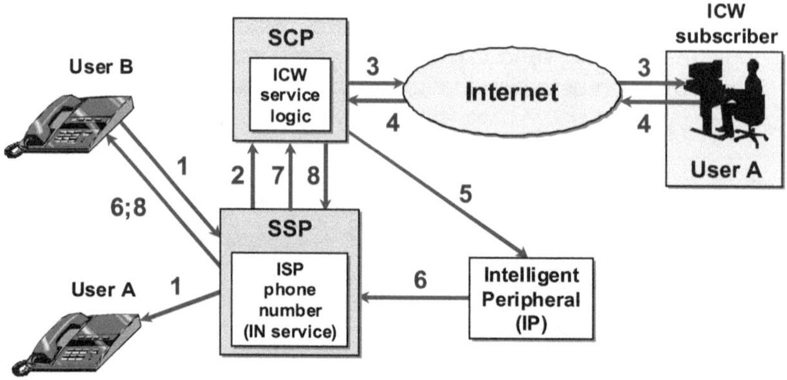

Figure 3.19 - ICW – Rejecting the call incoming.

Pre-conditions: User A is connected to Internet via a dial-up connection, an incoming call has arrived for User A, User A chooses to reject the call.

Post-conditions: The incoming call is terminated. The dial-up connection is still active.

- Step 1 - An user B invites the user A to a call.
- Step 2 - SSP sends request to SCP for service processing.
- Step 3 - SCP sends invitation message to user A through the Internet.
- Step 4 - The user A chooses to reject it.
- Step 5 - When the Internet user rejects the incoming call the ICW service logic requests to play an announcement to the user B saying that the callee has rejected the call.
- Step 6 - An announcement is played to the user B saying that the callee has rejected the call.
- Step 7 - When the announcement is finished the SSP notify the ICW service logic about the end of it.
- Step 8 - The ICW service logic then releases the call initiated by the user B.

3.6.2.6 ICW – Accepting the call on the phone

The user A is invited to a call by an user B and chooses to answer on his PSTN line (Figure 3.20).

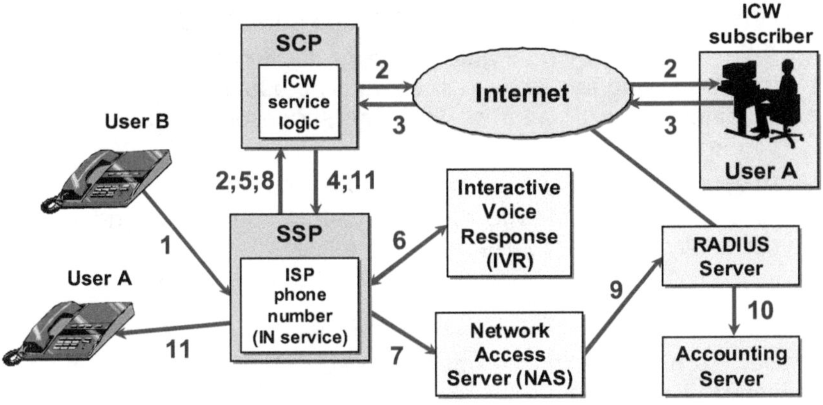

Figure 3.20 - ICW – Accepting the call on the phone.

Pre-conditions: The User A is connected to Internet by means of a dial-up connection, an incoming call has arrived for the User A, the User A chooses to answer the call on his PSTN line.

Post-conditions: The dial-up connection is terminated and the call is established between the User A and the User B using PSTN to PSTN connection.

- Step 1 - The user A is invited to a call by an user B.
- Step 2 - SCP sends invitation message to user A through the Internet .
- Step 3 - The ICW client software disconnects the dial-up connection.
- Step 4 - Therefore a disconnect signal is sent to the SSP. This is triggered by the SSP since an EDP9 (ODisconnect) related to the call between user A and the NAS was previously armed.
- Step 5 - Consequently the SSP sends an EventReportBCSM operation to the ICW service logic. The ICW service logic disarms TDP13 on user's A phone line by means of some management interface.
- Step 6 - Since a connection to an IVR was still established, the ICW service logic releases this connection. The SSP disconnect the connection to the IVR.
- Step 7 - And, sends to the NAS a message to disconnect it.

- Step 8 - As the result of the previous arming of an EDP9 (Odisconnect) related to the disconnection of the NAS the SSP sends an EventReportBCSM operation to the ICW service logic.
- Step 9 - The NAS detects the end of the dial-up connection and instructs the RADIUS gateway to stop the accounting.
- Step 10 - The RADIUS gateway asks the accounting server to stop the accounting for user's A account. The user's A profile is updated by deleting the IP address from the scope of the previous dial-up connection.
- Step 11 - The ICW service logic has just released the call to the NAS, therefore it tries to route the call to user's A phone line. The call finally is routed to user's A phone line.

3.6.2.7 ICW- Accepting the call on IP

The user A during his Internet session chooses to answer an incoming call using VoIP (Figure 3.21).

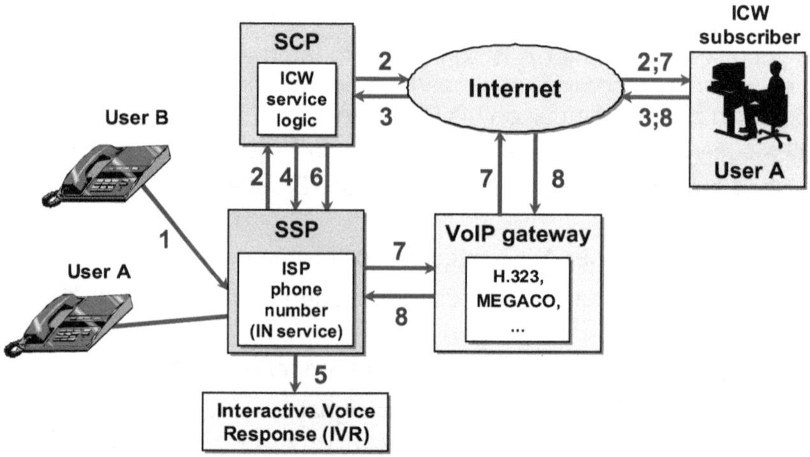

Figure 3.21 - ICW – Accepting the call on IP.

Pre-conditions: The User A is connected to Internet, an incoming call has arrived for User A, User A chooses to answer the call over IP.

Post-conditions: The call is established between User A and the User B using a VoIP gateway.

- Step 1 - The user A is connected to Internet, an incoming call has arrived for user A.

- Step 2 - SCP sends invitation message to user A through the Internet.
- Step 3 - User A chooses to answer the call over IP.
- Step 4 - Since a connection to an IVR was still established, the ICW service logic release this connection.
- Step 5 - The SSP disconnect the connection to the IVR.
- Step 6 - The ICW service logic retrieves User's A IP address and instructs the network to route the call to the appropriate VoIP gateway and requests to be notified of the outcome of the call routing.
- Step 7 - The SSP establish the PSTN connection to the VoIP gateway. The ICW service logic controls the VoIP gateway to terminate the call to user's A IP address. Depending on the VoIP gateway used, different control interfaces are possible e.g. H323, MeGaCo.
- Step 8 - The VoIP gateway terminates the call to user's A IP address.

3.6.2.8 Meeting Scheduler

The service is a meeting scheduler and initiator. A user has a web based user interface to schedule a meeting. He decides the time and the attendants. This information is recorded in the service profile, once it is decided it is authorized. The timing information also is used by an external application that launches the service.

The service retrieves the attendant's list from the service profile and uses the user profile to locate the users. Two kinds of communications are considered: VoIP and PSTN meaning that users can be located both on a telephone and a PC. The service establishes as many calls as needed and puts all of them in the same communication. Some of the information can reside within the user's domain (e.g., attendant list), being accessed from the service upon execution time.

Service features that have an impact on the PSTN-to-IP context are:

- Notification of phone attendance confirmation on meeting manager's PC - During the meeting confirmation phase the service logic checks for every meeting participant the availability of an e-mail address in his user profile. If the participant has no e-mail address, the service logic chooses a telephone number and a time at which the meeting participant might be available by elaborating the user profile data and uses this information to schedule, via an external application, an attendance confirmation via phone.

- The phone attendance confirmation process is activated by an external scheduler. Using the information related to a particular meeting attendant, a new call is created. The service logic instructs the call control component to interact with the service node to create a new call leg and to route it to the meeting participant's phone number. When he answers, he is requested to dial his PIN to authenticate him/her. After the successful authentication data check against his user profile information, a message is played informing him of the scheduled meeting and asking him to confirm his attendance. Meeting participant's answer (dialed digit) is then collected and depending on it, his attendance is confirmed in the meeting profile manager. The meeting manager is informed about this acceptance by means of a pop up window or by updating a graphi-cal application.
- Notification of meeting start on PC for VoIP users - After the meeting is scheduled the service informs the attendants about it for example by e-mail. Each potential attendant is requested to confirm his participation.
- The service can have two different choices:
    - It could check the user profile to get the location of the user at the time the meeting is scheduled.
    - It could ask the attendant to provide this information directly.
- In the first case an external scheduler to start a new conference triggers the meeting activation process. It has previously obtained a list of attendants from the service profile containing the identification of those users who have confirmed their attendance at the meeting. From the user profile, it has obtained a list of possible locations for each user, ordered by user preferences at the time when the conference is going to take place. Once it has obtained all the needed information related to the attendants, it starts the conference and notifies VoIP attendants of the meeting start by popping up a win-dow on their PC. They can at this point accept or deny to join the conference.
- Notification to the chair's PC of participants who left or joint the conference - During the execution of the service the conference status, in term of participants and time they have been logging in, is monitored. The chair (who scheduled the meeting) may be not log-ged in the conference. He is informed in any case about any change of the conference status. For example if a participant leaves or joint the conference the chair is notified about this on his/her PC and this information are stored in a file and used for billing off-line proce-dures.

- On-line notification which participant(s) is currently talking - The service could have the feature to update a graphical user interface, which maintains an image of the current conference status. So that participants who are also connected to Internet could receive graphical information about who is talking. This service feature depends on the capability of the resource, which provides the multiparty audio bridging feature to the service, to capture and use information about traffic flows for each leg.

3.6.2.9 Unified Communication - Interworking between different session e.g., telephone call and chat session

The unified communication service allows users to send, retrieve and receive messages disregarding the format and the terminal where the user is connected. The user should be able to create and respond to multimedia messages from any terminal and create and send any type of message without regard to the recipient's mailbox requirements. The user should also be able to reply to messages and forward messages with calls, and reply and forward calls with messages [Jha04].

Service features that have an impact on the PSTN-to-IP context are:

The messages are exchanged when different sessions (i.e., telephone call and chat session) are occurring (Figure 3.22).

- Step 1 - The two users (user A and user B) are involved in telephone call and two other users (user C and user D) are involved in chat session.
- Step 2 - Suddenly, user D knows that user's A mother is at the hospital. So, he sends a message (i.e., e-mail message) to user A marked as a high priority message.
- Step 3 - The provider notifies user A about the message.
- Step 4 - User A holds the phone call, and calls the messaging service.
- Step 5 - He receives the message in a voice format, text-to-speech conversion is required.
- Step 6 - The user A sends a reply in a voice form, the provider converts it in a text format and sends an e-mail to user D.
- Step 7 - Meanwhile user A unholds the call to inform user B he has to hang-up.

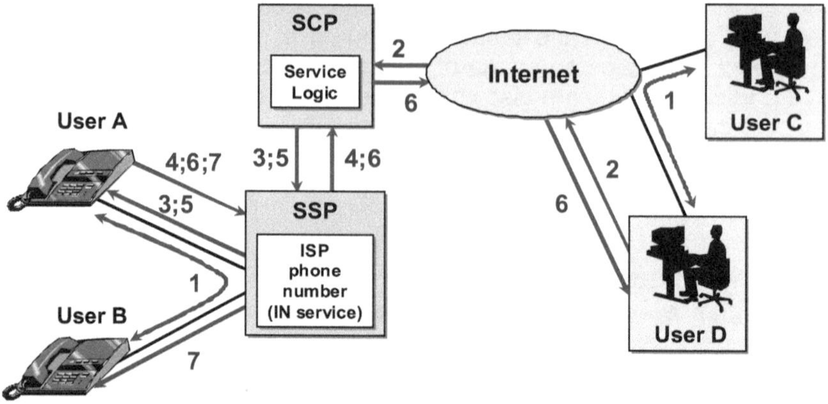

Figure 3.22 - Interworking between different sessions.

3.6.2.10 Unified Communication - Distributed and Enhanced Call Center

The distributed and enhanced call center scenario shows several sides of IN-Internet integration. It features the speech connection of a PSTN terminal to a VoIP terminal, where the service logic controlling the call resides in a third party domain. It also shows how a call between a VoIP terminal and a PSTN terminal could be set up initiated from the Internet, through a third party service. The call center agents are provided with a PC with a web interface. The PC is connected to Internet, and it uses this connection to register/deregister availability of handling incoming calls (data communication), and to establish the communication with the client using VoIP (voice communication). This solution has the advantage that the call center can have the agent positions totally distributed and the agent could even be located in his own home. The agent is also able to place outgoing calls by the use of a click-to-dial like service to initiate a call (e.g., PC-to-phone). The logic and data for selecting the "best" available agent is allocated in an external domain seen from the IN point of view. It could reside in the company administration domain or this service could be outsourced to another company. The logic and data for agent selection could also be allocated in the IN domain, but this case is not examined further here.

Service features that have an impact on the PSTN-to-IP context are: - Incoming call notification and advanced reply.

The best available agent is notified on his PC about incoming calls directed to him. He is presented with a menu window, which allows him to choose between the following choice:

- Accept the call,

- Reject the call,
- Forward the call to another agent,
- Reply by typing a message that is translated into speech,
- Reply by sending back a dynamic web page built by composing information.

3.6.3 IN Service Request from IP Networks

It must be possible to open the access to IN functions during an internet session. It enables:

- PINT-like services: click-to-dial, click-to-fax and voice access to web content,
- Access to customer and service information,
- Access to IN routing services, IP telephony gateway using IN-based address resolution.

The services may required specific intelligent peripheral resources with media transformation like text-to-fax and text-to-speech. The following services can well help in defining requirements in the context of how services supported by IN entities can be started from Internet requests.

3.6.3.1 Click-to-Dial (C2D) - Invocation

A user is able to initiate a telephone call by clicking a button during a web session. The called address (as well as the caller address) is either an IP address or a phone line number. The charging party could be either the initiator or one of the called parties.

A user is able to initiate a telephone call by clicking a button during a web session (Figure 3.23).

Pre-conditions: The user is logged on to the system and has selected the click-2-dial service for invocation.

Post-conditions: A call has been established between the origination and the destination.

- Step 1 - The user invokes the Click2Dial service.
- Step 2 - The service logic contacts call control to setup a call to setup a call between the caller and the called user. It must be possible for a user to initiate a call between two parties, neither of which is the user. If the user invoking the call is to be billed, this means that the C2D service has to known the initiator of the call so that the information may be passed to billing system. In the C2D service, especially if the call is placed by a third party, it is difficult to determine

who the originating party is and who the destination party is. This has an impact on the use of the Parlay interface insofar as some order must be placed on the parties so that there is an origination and a destination to the call. The originating address is taken as the first address entered (user A) and the destination address is the se-cond address entered (user B). The user (this is the user to be billed for the service) requests that a call be placed between two users: user A and user B. The C2D service is requested to initiate a call between the indicated parties.

- Step 3 - The C2D first initiates the routing of the call towards the first party in the call (user A).
- Step 4 - The C2D service is notified that the routing of the call towards the first party has been successful.
- Step 5 - Then it initiates the routing of the call towards the second party in the call (user B) and it is notified that the routing of the call towards the called party has been successful.
- Step 6 - The call now is active.

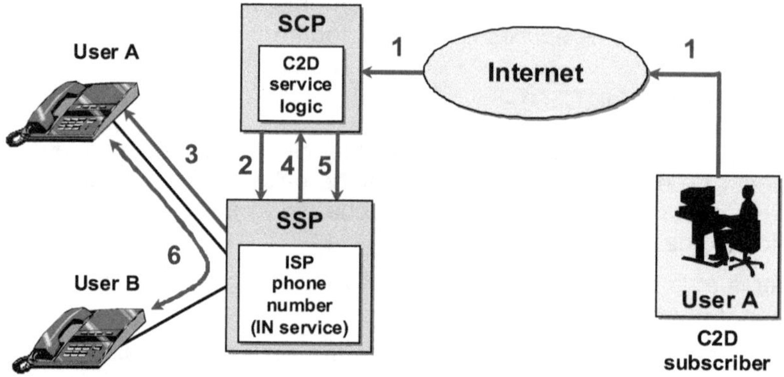

Figure 3.23 - Click2Dial invocation.

3.6.3.2 C2D - Call setup: PSTN as originating party terminal

A user initiates a telephone call by clicking a button during a web session (Figure 3.24).

Pre-conditions: The calling user is logged on to the system and has selected/invoked the click-to-dial service.

Post-conditions: The call has been set up towards the origination address. A call link has been established towards the origination address.

- Step 1 - The calling user has selected the click-to-dial service.
- Step 2 - The address of the origination is resolved from the user profile. If the originating user is registered on a PSTN phone.
- Step 3 - Then, the originating call link is routed towards the PSTN.
- Step 4 - First the user's A user profile is contacted to return the correct terminal to which the call must be delivered: in this case a PSTN address.
- Step 5 - An invocation to create a new call is sent to the call control manager.
- Step 6 - Which returns the call session ID. The call manager creates a new call.
- Step 7 - At this point the service logic initiates the routing of the call towards the first party in the call (user A) and waits the result of this operation.
- Step 8 - The result is sent to the C2D service logic.

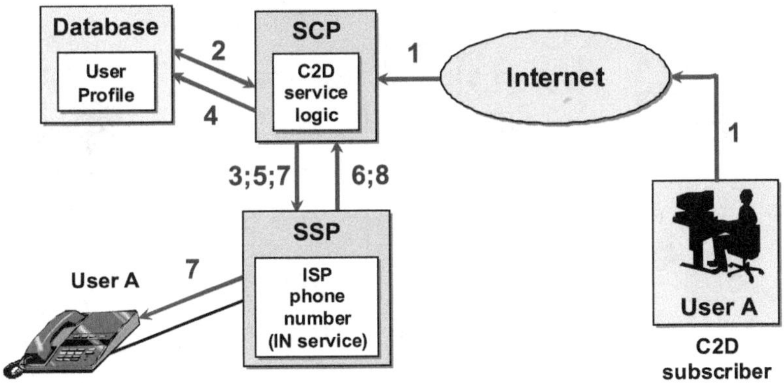

Figure 3.24 - Call setup: PSTN as originating party terminal.

3.6.3.3 C2D - Call setup: PSTN as destination party terminal (PSTN-PSTN)

A user initiates a telephone call by clicking a button during a web session (Figure 3.25).

Pre-conditions: The call has been routed to the origination.

Post-conditions: The call has been set up towards the destination. A call link has been established towards the destination. The call setup is complete.

- Step 1 - The calling user has selected the click-to-dial service.
- Step 2 - The address of the destination is resolved from the user profile. The user is registered on a PSTN.
- Step 3 - The call is routed towards the destination.
- Step 4 - The C2D service logic requests to user's B user profile to return the correct terminal to which the call should be delivered. The address of the terminal to which the call should be delivered - in this case a PSTN address - is returned.
- Step 5 - The C2D service logic initiates the routing of the call towards the second party in the call (user B).
- Step 6 - The result of this operation is sent to the C2D service logic.
- Step 7 - The call setup is complete.

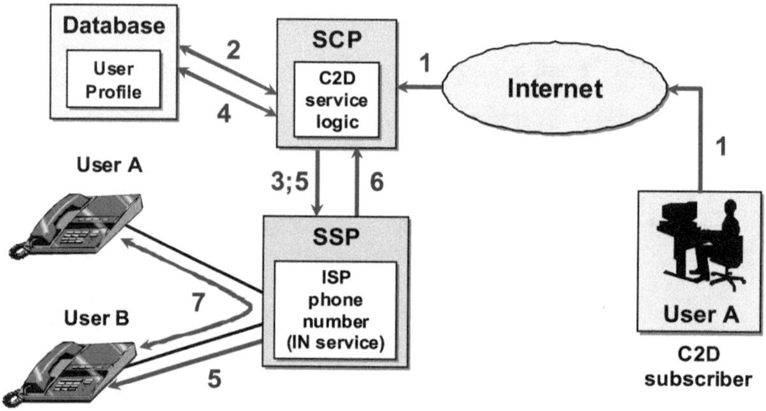

Figure 3.25 - Call setup: PSTN as destination party terminal (PSTN-PSTN).

3.6.3.4C2D - Call setup: PC as destination party terminal (PSTN-PC)

A user initiates a telephone call from PSTN to PC (Figure 3.26).

Pre-conditions: A call link has been established to the origination.

Post-conditions: The call has been set up towards the destination. A call link has been established towards the destination. The call setup is complete.

- Step 1 - A call link has been established to the origination.
- Step 2 - The address of the destination (user B) is resolved from the user profile. The user is registered on a PC. The call is routed towards the destination.

- Step 3 - The C2D service logic requests to user's B user profile to return the correct terminal to which the call should be delivered. The address of the terminal to which the call should be delivered - in this case an IP address - is returned.
- Step 4 - The service logic then invites user B to the current session by using the user ID of the user and the address of the terminal at which the user is to be invited. SIP is used as the invitation mechanism.
- Step 5 - User responses to the invitation.
- Step 6 - The response to the invitation is sent to the service logic.
- Step 7 - In case the user accepted, the C2D service logic initiates the routing of the call towards the second party in the call (user B).
- Step 8 - The result of this operation is sent to the C2D service logic.

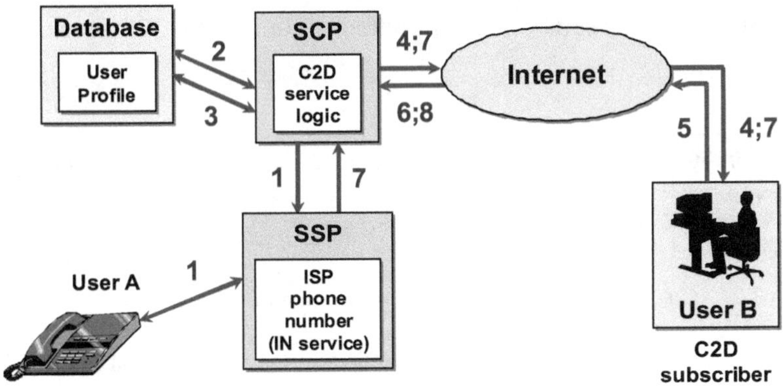

Figure 3.26 - Call setup: PC as destination party terminal (PSTN-PC).

3.6.3.5 C2D - Call setup: PC as originating party terminal

A user initiates a telephone call by clicking a button during a web session (Figure 3.27).

Pre-conditions: The calling user is logged on to the system and has selected/invoked the click-to-dial service.

Post-conditions: The call has been set up towards the origination.

- Step 1 - The calling user has selected the click-to-dial service.
- Step 2 - The address of the origination is resolved from the user profile. If the originating user is registered on a PC.

- Step 3 - Then, a call link is routed towards the Internet. In reality the call is not routed yet. But from the point of view of the service logic, the call control behaves as if the call has been routed correctly.
- Step 4 - The C2D service logic requests to user's A user profile to return the correct terminal to which the call should be delivered. The address of the terminal to which the call should be delivered - in this case an IP address - is returned.
- Step 5 - The service logic then invites user A to establish a new session. SIP is used as the invitation mechanism.
- Step 6 - User responses to the invitation.
- Step 7 - The response to the invitation is sent to service logic. In case user A accepts service logic instructs the call manager to create a new call. The C2D service logic initiates the routing of the call towards the first party in the call (user A).
- Step 8 - The result of this operation is sent to the C2D service logic.

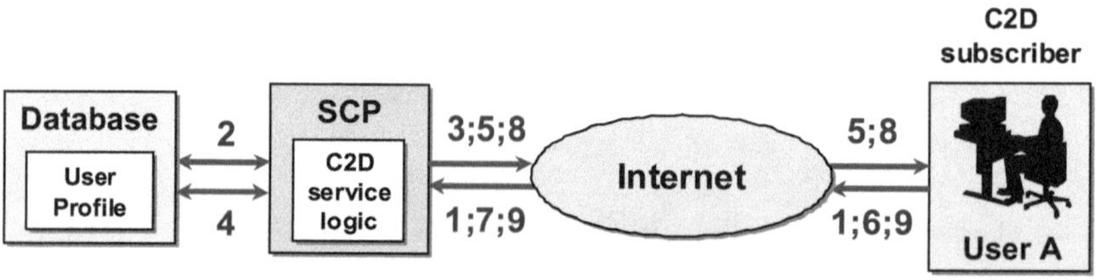

Figure 3.27 - Call setup: PC as originating party terminal.

3.6.3.6 C2D - Call setup: PSTN as destination party terminal (PC-PSTN)

A user initiates a telephone call by clicking a button during a web session (Figure 3.28).

Pre-conditions: The call has been routed to the origination.

Post-conditions: The call has been set up towards the destination. A call link has been established towards the destination. The call setup is complete.

- Step 1 - The calling user has selected the click-to-dial service.
- Step 2 - The address of the destination is resolved from the user profile. The user is registered on a PSTN.
- Step 3 - The call is routed towards the destination.
- Step 4 - The C2D service logic requests to user's B user profile to return the correct terminal to which the call should be delivered. The

address of the terminal to which the call should be delivered - in this case a PSTN address - is returned.

- Step 5 - The C2D service logic initiates the routing of the call towards the second party in the call (user B).
- Step 6 - The result of this operation is sent to the C2D service logic.
- Step 7 - The call setup is complete.

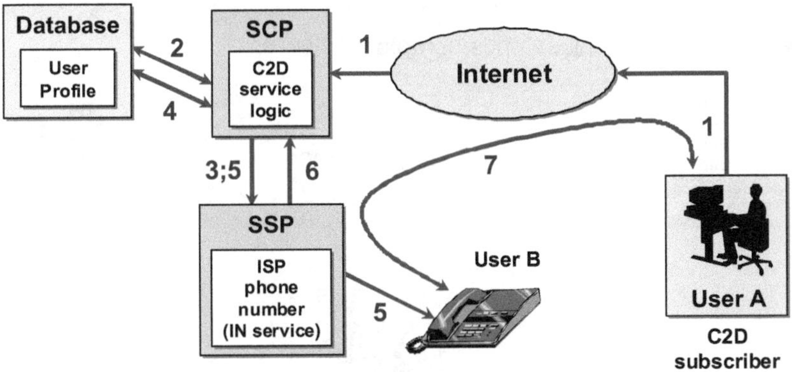

Figure 3.28 - Call setup: PSTN as destination party terminal (PC-PSTN).

3.6.3.7 C2D - Call setup: PC as destination party terminal (PC-PC)

A user initiates a call by clicking a button during a web session (Figure 3.29).

Pre-conditions: A call link has been established to the origination.

Post-conditions: The call has been set up towards the destination. A call link has been established towards the destination. The call setup is complete.

- Step 1 - The calling user has selected the click-to-dial service.
- Step 2 - The address of the destination is resolved from the user profile. The user is registered on a PC. The call is routed towards the destination.
- Step 3 - The C2D service logic requests to user's B user profile to return the correct terminal to which the call should be delivered. The address of the terminal to which the call should be delivered - in this case an IP address - is returned.
- Step 4 - The service logic then invites user B to the current session by using the user ID of the user and the address of the terminal at

which the user is to be invited. SIP is used as the invitation mechanism.

- Step 5 - User responses to the invitation.
- Step 6 - The response to the invitation is sent to the service logic.
- Step 7 - In case the user accepted, the C2D service logic initiates the routing of the call towards the second party in the call (user B).
- Step 8 - The result of this operation is sent to the C2D service logic.
- Step 9 - The call setup is complete [Bla00].

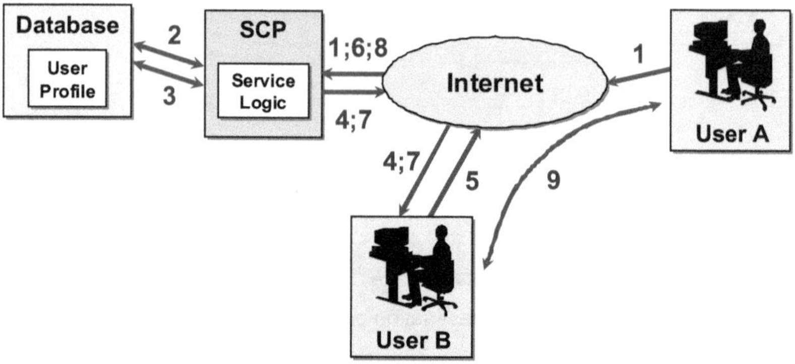

Figure 3.29 - Call setup: PC as destination party terminal (PC-PC).

3.6.3.8 Meeting Scheduler (MS)

A user has a web based user interface to schedule a meeting (Figure 3.30). He will decide the time and the attendants. This information is recorded for the service profile, once it is decided it is authorised. The timing information is also used by an external application that launches the service via the access function. Every meeting attendant is confirmed, either with an e-mail notification and a following confirmation process via a web page or with a phone call. Once the service is launched, it uses the attendants list and the service profile to locate the users.

Service features that have an impact on the IP-to-PSTN context are: - Starting a meeting by selecting a user(s) and timing on a Web page.

Two kinds of communications are considered: VoIP and PSTN. The service sets up the communication to the end users using a service node. The service node has a direct connection to the PSTN and uses the call control component, controlling the vocal gateway to establish the connection to the VoIP terminals. The service establishes as many calls as needed and puts all of them in the same communication.

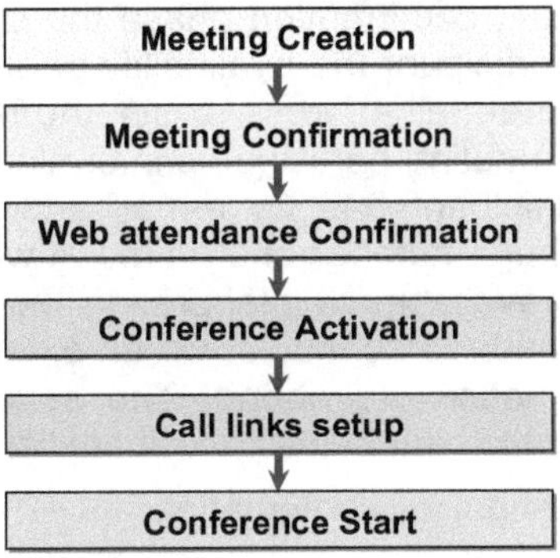

Figure 3.30 - The meeting schedule algorithm.

The algorithm steps description:

- Step 1 - Meeting creation.

  Pre-conditions: The Meeting manager must have a web access to log in the system and to set the MS service.

  Post-conditions: Meeting confirmation process is started.

  The meeting manager accesses via a web browser the provider web page containing the MS applet. He sets relevant information for the meeting to take place, e.g. meeting attendants and date/time of the meeting - along with some authentication information to confirm meeting manager's identity.

  The web server launches the meeting scheduler servlet, which interacts with the service logic. The MS servlet checks in order to confirm the meeting manager authentication data. Furthermore it checks if the meeting participants' identifier does really exist and whether the terminals specified in the user profile configurations meet the service requirements. The MS servlet then, accesses the meeting profile manager to include the new meeting information.

  Meeting profile manager returns a meeting identifier which has to be used by the meeting participants to confirm their attendance to the meeting. This identifier is also shown to the meeting manager in a HTML page confirming the success of the operation.

- Step 2 - Meeting confirmation.

  Pre-conditions: A meeting must have been created following the meeting start process.

  Post-conditions: Participants are informed about the meeting by e-mail or by phone.

During the meeting confirmation phase the service logic checks for every meeting participant the availability of an e-mail address in his user profile. If an e-mail address exists, the MS service logic sends an e-mail to the meeting participants informing him of the meeting. If the meeting participant has no e-mail address, the service logic chooses a telephone number and a time at which the meeting participant might be available by elaborating the user profile data and uses this information to schedule, via an external application, an attendance confirmation via phone. The sequence is repeated for every meeting participant in the participant list of the meeting.

- Step 3 - Web attendance confirmation.

  Pre-conditions: The participant received an e-mail including the meeting attendance confirmation URL, meeting identifier and related meeting information like the participant list.

  Post-conditions: The participant confirms his attendance.

  The participant accesses via a web browser the web page containing the meeting attendance confirmation form. The participant is authenticated by means of user name and password which are checked against personal data in his user profile. Then he selects the confirmation choice and finally he is shown with a HTML page to notify the success of the operation.

- Step 4 - Conference activation.

  Pre-conditions: The meeting activation process has obtained the list of possible points of presence for of the participants.

  Post-Conditions: The conference is activated.

  The meeting activation process is activated by a time-dependent scheduler to start a new conference. It has previously obtained a list of attendants from the meeting profile manager containing the information related to those users who confirmed their attendance.

  From the user profile, it gets a list of possible locations for each user, ordered by user preferences at the time when the conference is going to take place.

  Once it has obtained all the needed information related to the attendants, the service logic creates a new call instance in the service node through interacting with the call control.

- Step 5 - Call links set up.

  Pre-conditions: A new call has been created but no call link is attached yet.

Post-conditions: The participants' call links are created

Once the conference has been started and a call has been created, the meeting activation process needs to create a call link for each of the attendants.

The first point of presence of the first attendant is selected and the call control is asked to route the call link to that address. If the address is a telephone number, the call control just instructs the service node to route the call link to it. If the address is an IP address, the call link has to be routed through a vocal gateway.

Consequently the call control instructs the service node to route the call link to one of the phone numbers assigned to the vocal gateway and instructs the gatekeeper to translate that number to the desired IP address. When the call reaches the vocal gateway it requires from the gatekeeper, using RAS protocol, the target IP address. If the end user does not answer the call, the next point of presence is selected and the call link is routed to this new address. This procedure is repeated until the end user answers or all possible addresses are tried. When a user answers, music or some announcement is played to him until the process to call all the attendants is finished. At that point the conference is ready to be set up.

- Step 6 - Conference start.

  Pre-conditions: A call has been created and all the related call links have also been created, but they are not attached yet.

  Post-conditions: The conference has been completely set up and the attendants can talk to each other.

  When all attendants have been contacted and they have answered (or at least all their different possible addresses have been tried) the conference can be started. As a first step the music/announcement is stopped. Then the service node is instructed to create a conference and to attach the call links to it [Bla00, Lao05].

3.7 PINT and SPIRITS Protocols for Converged Services Support

3.7.1 PINT Protocol

The desire to invoke certain telephone call services from the Internet has been identified by many different groups (users, public and private network operators, call center service providers, and equipment vendors). The generic scenario is as follows (when the invocation is successful):

- an IP host sends a request to a server on an IP network;

- the server relays the request into a telephone network;
- the telephone network performs the requested call service.

As examples, it is taken into consideration a user who wishes to have a callback placed to his telephone. It may be that a customer wants someone in the support department of some business to call them back. Similarly, a user may want to hear some announcement of a weather warning sent from a remote automatic weather service in the event of a storm.

The term "PSTN/Internet Interworking (PINT) service" denote such a complete transaction, starting with the sending of a request from an IP client and including the telephone call itself. PINT services are distinguished by the fact that they always involve two separate networks:

> an IP network to request the placement of a call, and the global switched telephone network (GSTN) to execute the actual call. It is understood that intelligent network systems, private PBXs, mobile communication networks, and the ISDN can all be used to deliver PINT services. Also, the request for service might come from within a private IP network that is disconnected from the whole Internet.

The requirements for the PINT protocol were deliberately restricted to providing the ability to invoke a small number of fixed telephone call services. Great care has been taken, however, to develop a protocol that is aligned with other Internet protocols where possible, so that future extensions to PINT could develop along with Internet conferencing.

Within the Internet conference architecture, establishing media calls is done via a combination of protocols. SIP is used to establish the association between the participants within the call. This association between participants within the call is called a "session". And, SDP is used to describe the media to be exchanged within the session. The PINT protocol uses these two protocols together, providing some extensions and enhancements to enable SIP clients and servers to become PINT clients and servers.

A PINT user who wishes to invoke a service within the telephone network uses SIP to invite a remote PINT server into a session. The invitation contains an SDP description of the media session that the user would like to take place. This might be a "sending a fax session" or a "telephone call session", for example. In a PINT service execution session, the media is transported over the phone system, while in a SIP session the media is normally transported over an internet.

When used to invoke a PINT service, SIP establishes an association between a requesting PINT client and the PINT server that is responsible for invoking the service within the telephone network. These two entities are not the same entities as the telephone network entities involved in the telephone network service. The SIP messages carry within their SDP payloads a description of the telephone network media session.

A PINT server accepts an invitation and a session is established is no guarantee that the media will be successfully transported. This is analogous to the fact that if a SIP invitation is accepted successfully, this is no guarantee against a subsequent failure of audio hardware.

The particular requirements of PINT users lead to some new messages. When a PINT server agrees to send a fax to telephone B, it may be that the fax transmission fails after part of the fax is sent. Therefore, the PINT client may wish to receive information about the status of the actual telephone call session that was invoked as a result of the established PINT session. Three new requests, SUBSCRIBE, UNSUBSCRIBE, and NOTIFY, are added here to the basic SIP to allow this.

The enhancements and additions specified here are not intended to alter the behavior of baseline SIP or SDP in any way. The purpose of PINT extensions is to extend the usual SIP/SDP services to the telephone world. Apart from integrating well into existing protocols and architectures, and the advantages of reuse, this means that the protocol specified here can handle a rather wider class of call services than just the milestone services.

3.7.1.1 PINT Milestone Services

The original motivation for defining this protocol was the desire to invoke the following three telephone network services from within an IP network:

- Request-to-call - A request is sent from an IP host that causes a phone call to be made, connecting party A to some remote party B.
- Request-to-fax-content - A request is sent from an IP host that causes a fax to be sent to fax machine B. The request MAY contain a pointer to the fax data that could reside in the IP network or in the telephone Network, OR the fax data itself. The content of the fax MAY be text OR some other more general image data. The details of the fax transmission are not accessible to the IP network, but remain entirely within the telephone network. This service does not relate to fax-over-IP: the IP network is only used to send the request that a certain fax be sent. Of course, it is possible that the resulting telephone network fax call happens to use a real-time IP fax solution, but this is completely transparent to the PINT transaction.
- Request-to-speak/send/play-content - A request is sent from an IP host that causes a phone call to be made to user A, and for some

sort of content to be spoken out. The request MUST EITHER contain a URL pointing to the content, OR include the content itself. The content MAY be text OR some other more general application data. The details of the content transmission are not accessible to the IP network, but remain entirely within the telephone network. This service could equally be called request-to-hear-content; the user's goal is to hear the content spoken to them.

3.7.1.2 Relation between PINT Milestone Services and Traditional Telephone Services

There are many different versions and variations of each telephone call service invoked by a PINT request. There may be thousands of agents in the call center, and there may be any number of sophisticated algorithms and pieces of equipment that are used to decide exactly which agent will return the call. And once this choice is made, there may be many different ways to set up the call: the agent's phone might ring first, and only then the original user will be called; or perhaps the user might be called first, and hear some horrible music or pre-recorded message while the agent is located.

Similarly, when a PINT request causes a fax to be sent, there are hundreds of fax protocol details to be negotiated, as well as transmission details within the telephone networks used.

PINT requests do not specify too precisely the exact telephone-side service. Operational details of individual events within the telephone network that executes the request are outside the scope of PINT. This does not preclude certain high-level details of the telephone network session from being expressed within a PINT request.

For example, it is possible to use the SDP "lang" attribute to express a language preference for the request-to-hear-content service. If a particular PINT system wishes to allow requests to contain details of the telephone-network-side service, it uses the SDP attribute mechanism.

3.7.1.3 PINT Functional Architecture

PINT clients and servers are SIP clients and servers. SIP is used to carry the request over the IP network to the correct PINT server in a secure and reliable manner, and SDP is used to describe the telephone network session that is to be invoked or whose status is to be returned.

A PINT system uses SIP proxy servers and redirect servers for their usual purpose, but at some point there must be a PINT server with the means to relay received requests into a telephone system and to receive acknowledgement of these relayed requests. A PINT server with this capability is called a "PINT gateway". A PINT gateway appears to a SIP system as a user agent server. Notice that a PINT gateway appears to the PINT infrastructure as user, while in fact it really represents an entire telephone network infrastructure that can provide a set of telephone network services.

The PINT system might appear to an individual PINT client shown in Figure 3.31:

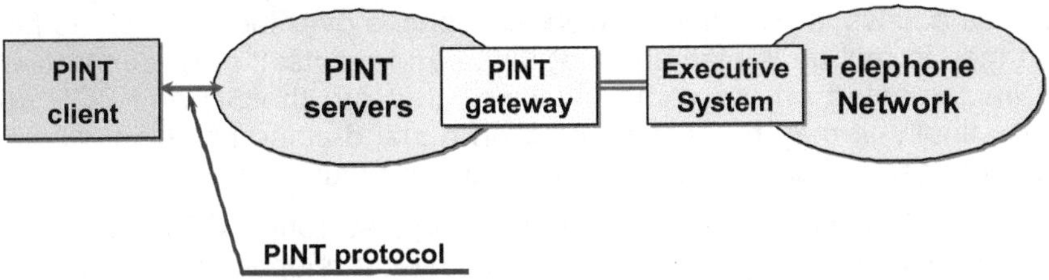

Figure 3.31 - PINT Functional Architecture.

A single PINT request might pass through a series of location servers, proxy servers, and redirect servers, before finally reaching the final PINT gateway that can actually process the request by passing it to the telephone network cloud.

The PINT gateway might have a true telephone network interface, or it might be connected via some other protocol or API to an executive system that is capable of invoking services within the telephone network.

As an example, within an intelligent network, the PINT gateway might appear to realise the service control gateway function. In an office environment, it might be a server adjunct to the office PBX, connected to both the office LAN and the office PBX.

The executive system that lies beyond the PINT gateway is outside the scope of PINT [RFC 2848].

3.7.2 SPIRITS Protocol

- Provides the relevant background explaining the mechanism for interactions between the PSTN and SPIRITS Server.
- Based on this mechanism, provides basic requirements and sets forth the methodology for constructing the building blocks of the SPIRITS protocol.
- Provides particular examples of application of this methodology regarding the leading SPIRITS benchmark service - the Internet call waiting (ICW).

3.7.2.1 Physical Architecture

Figure 3.32 depicts the joint PSTN/Internet physical architecture relevant to SPIRITS. The services are invoked and, subsequently, the SPIRITS protocol is initiated when a message from a SPIRIT client located in the IN service control point (SCP) or service node (SN) arrives either, on interface E or A, respectively to the SPIRITS Server. Both E and A interfaces are over the IP network (very likely, over the Internet). The SPIRITS Server has access to PCs and network appliances over the Internet. Its function may in fact be implemented at these end-points; it may, alternatively, be implemented on the same machines that run the SN and SCP software; finally, it may be implemented on a stand-alone machine, which is the most general case and is thus depicted in Figure 3.32.

In most practically important cases, the request from a SPIRITS client is caused by a request from a central office (i.e., a telephone switch) sent to either the SCP or SN, although the Internet-based service initiation by these elements that had not been triggered by the central office is theoretically possible. Definitely, none of the SPIRITS benchmark services are initiated in such a way, so for the purposes of the SPIRITS protocol development, it should be assumed that the service invocation was a direct result of an earlier action by the central office.

When the central office realizes that an external action is needed it either proceeds with issuing a query to the SCP, to which it has no bearer circuit connection, only a signaling one, over SS7 or forwards the call to the SN, to which it has the ISDN connection.

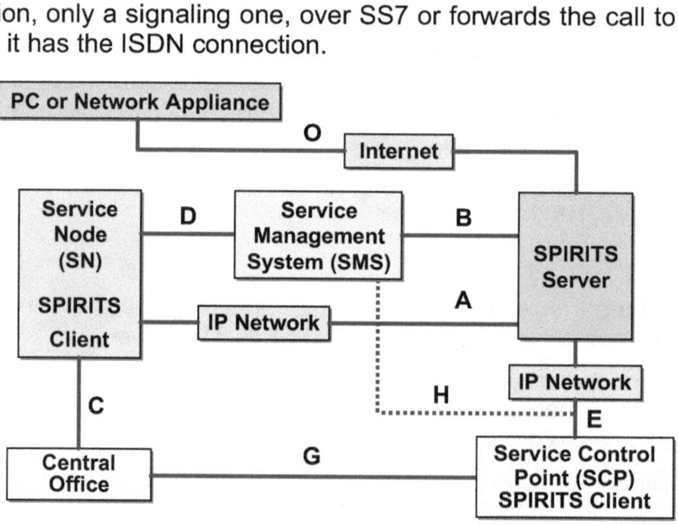

Figure 3.32 - SPIRITS Physical Architecture.

The SCP and SN are described in the ITU-T Recommendations Q.1205, Q.1215, and Q.1225. The present SN implementations may contain both the signaling system No.7 and ISDN access interfaces, so that such SNs may act as pure SCPs. Still for the purposes of a particular SPIRITS service instance, the SN would act either as an SCP or original SN.

The architecture excludes direct communications of the Central Office with the SPIRITS server. Such communications can be achieved only by means of either the SCP or SN. Moreover, as far as SPIRITS is concerned, there must be no difference in the protocol whether the connection is with the SCP or SN on the PSTN side.

3.7.2.2 The Role of the IN Call Model

The central office determines that it needs an external action based on its call model, which does not necessarily have to be the IN call model. If the central office does not use the IN means directly, it may establish a circuit to the SN; the latter then picks up the IN processing. Alternatively, if the Central Office does use the IN basic call state model (BCSM), it may issue the request to the SCP when the external processing is needed.

BCSM is standardized in the ITU-T Recommendation Q.1204 (general aspects), Q.1214 (IN capability set 1 aspects), and Q.1224 (IN capability set 2 aspects). The model guides all the aspects of the service request initiation. Because neither the internal call models (i.e., the ones that support switch-based services) nor the internal SN interfaces (i.e., inter-faces between the call processing and service processing) are standar-dized, the SPIRITS protocol can be influenced normatively only by the BCSM and the protocol, intelligent network application part (INAP), which accompanies it.

Of course, neither the BCSM, nor INAP are subject to standardization in SPIRITS; however, both have direct influence on SPIRITS in as much as the former effectively defines the service initiation events and the corresponding minimum set of messages that are initially issued by the SPIRITS client and the latter defines what information is available at the SPIRITS client at the time the service is initiated. In fact, the same set of events can re-occur later in the service, and, consequently same messages can be sent in the middle of the service-supporting exchange.

The BCSM specifies two finite state machines: one for the originating, and one for the terminating part of the call. In addition to a call state, called point in call (PIC), each part also specifies special states called detection points (DPs) that are associated with the transitions between PICs. The PICs are best viewed as primary states in that DPs are directly associated with the transitions from PIC to PIC. Thus, the basic construct of the

BCSM is PIC>DP>PIC, although non-DP-associated transitions (i.e., PIC>PIC) also exist.

If a transition passes through a DP, the central office pauses and examines:

- whether a DP is armed (i.e., active), and
- whether it meets the criteria for launching an IN query or sending a notification.

If a DP is armed (off-line) from the service management system (SMS), it is called a trigger. The trigger is static in that it is armed forever. All SPRITS services are initiated by triggers. But the same DPs that can be set off-line as triggers can also be set on line - for the duration of a particular call and only for that call - by the SCP. Regardless of whether they are armed statically or dynamically, DPs can be armed either as notifications or requests. Correspondingly, depending on the type of the active DP, the central office can either issue a notification (and transit to the next PIC) or, suspend call processing, issue a request to the SCP, and wait for the response. When the response comes back, it may contain an instruction on how to continue with the call. Such an instruction may even break the model by causing a "go-to"-type transition to any other PIC. The use of SMS can sometimes be avoided by implementing its service distribution and trigger-setting function by other means. The SMS thus is included here for completeness.

On the PSTN side, the purposes of the Internet call waiting (ICW) can be met with one of the two approaches - one based on the SN, and one on the BCSM-to-service control (i.e., central office-to-SCP) interaction. With the SN-based approach, when the central office detects that the called party is busy, it simply forwards the call to the SN, whose service logic initiates the message exchange by sending the notification to the SPIRITS server. With the SCP-based approach, the central office can use either of the two DPs of the terminating BCSM: "termination attempt authorized" or busy. Either DP, of course, would need to be armed as trigger. Nevertheless, the SPIRITS server must see the same notification, independent of the approach.

Once the interactions between the SPIRITS client and SPIRITS server started, the latter can instruct the client to arm additional DPs for the duration of the call. The client, could, in turn, send appropriate messages to the central office for the SCP-based implementation or enable appropriate internal events in an implementation-dependent manner in the SN for the SN-based implementations [Fay00].

3.7.2.3 Brief Description of Example SPIRITS Services

It is given a brief description of the example SPIRITS services to illustrate the motivation for the overall SPIRIT architecture:

- Internet call waiting (ICW),
- Internet caller-ID delivery, and
- Internet call forwarding.

These services are considered from the end-user point of view under the assumptions below:

- Service subscription or cancellation is a separate process and may be done over the telephone, via postal mail, or over the Web.
- The subscriber's IP host (e.g., a PC) is loaded with the necessary software including a personal identification number (PIN) and the IP addresses of the SPIRITS servers for realizing the SPIRITS services. The software may be sent by postal mail or downloaded from the web.
- The subscriber activates a SPIRITS service by an act of service session registration, which can take place anytime after he is connected to the Internet. The subscriber may specify the life span of the session. As soon as the session ends, the SPIRITS service is deactivated. Naturally, the subscriber should also be able to deactivate a SPIRITS service anytime during the service session.

3.7.2.4 SPIRITS Architecture

Figure 3.33 depicts the SPIRITS architecture, which includes the following entities:

- Service Control Function (SCF) - executes service logic, interacts with the entities in the IP domain (e.g., the SPIRITS proxy and PINT server) through the SPIRITS client, and instructs the switches on how to complete a call. Physically, the SCF may be located in either stand-alone general-purpose computers called service control points (SCPs) or specialized pieces of equipment called service nodes (SNs).
- Service Switching Function (SSF) - normally resides in a switch and is responsible for the recognition of intelligent network (IN) triggers and interactions with the SCF.
- SPIRITS Client - responsible for receiving PSTN requests from the SCF as well as sending responses back. It may be co-located with the SCF. If not, it communicates with the SCF over the D interface.

- PINT Server - receives PINT requests from the PINT client and relays them to the PSTN for execution over the E interface.
- SPIRITS Proxy - co-located with the PINT server and serves as an intermediary between the SPIRITS server and SPRITS client via the B and C interfaces, respectively.
- PINT Client - resides in the subscriber's IP host and is responsible for initiating PINT requests, which are sent to the PINT server over the A interface.
- SPIRITS Server - terminates PSTN requests and is responsible for all interactions (e.g., incoming call notification and relaying the call treatment) between the subscriber and the SPIRITS proxy.

Interfaces between the entities are as follows:

- Interface A - for sending PINT request to PINT Server. Its principal use is for service session registration and as a result activation of a SPIRITS service. In addition, this interface may be used for service subscription.
- Interface B - serves two main purposes:
    1. to notify the subscriber of incoming calls together with the calling number and name, if available; and
    2. to send to the SPRITS Proxy the subscriber's choice of call disposition specified on the fly.
- Interface C - for communications between the SPIRITS client and SPIRITS proxy. The SPIRITS proxy may in turn communicate with the SPRITS Server, or may act as a virtual server, terminating the requests without sending them down to the SPIRITS server.
- Interface D - for communications between the SPIRITS client and the SCF. Specifically, from the SCF to the SPIRITS client, the parameters associated with the applicable IN triggers are sent. From the SPIRITS client to SCF, the subscriber's call disposition is sent. The SCF transforms the user's disposition into appropriate actions, such as playing an announcement to the caller, and resuming the suspended call processing in the SSP.
- Interface E - for sending PINT requests to the SCF for execution [Fay01].

3.7.3 PINT and SPIRITS Protocols Mirror Images of Integration

The joint PINT/SPIRITS architecture is depicted in Figure 3.33. It is assumed that the SPIRITS client is either co-located with the IN service control function (SCF) or communicates with it over the PSTN-specific interface D in such a way so as to act on behalf of the PSTN/IN.

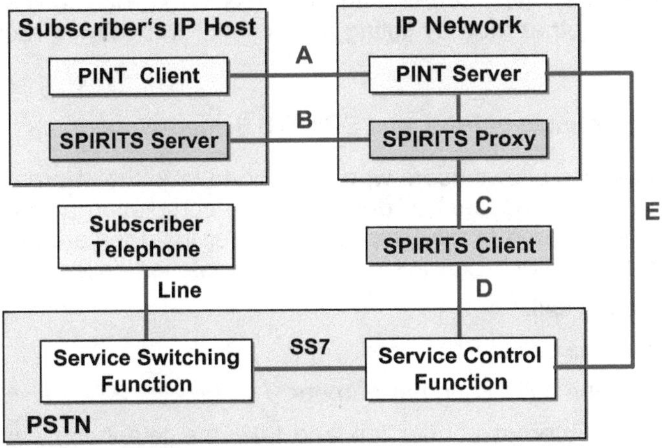

Figure 3.33 - Joint PINT/SPIRITS Architecture.

The SPIRITS services are invoked and, subsequently, the SPIRITS protocol is initiated when a message from a SPIRITS client, located in the IN service control point (SCP) or service node (SN) arrives on interface C to the SPIRITS proxy. The SPIRITS proxy processes the message and, in turn, passes it on over the interface B to the SPIRITS server. In most practically important cases, the request from a SPIRITS client is ultimately caused by a request from a central office (i.e., a telephone switch) sent to either the SCP or SN, although the Internet-based service initiation by these elements that had not been triggered by the central office is theoretically possible. Definitely, none of the SPIRITS benchmark services are initiated in such a way, so for the purposes of the SPIRITS protocol development, it should be assumed that the service invocation was a direct result of an earlier action by the central office.

With PINT, and that also applies to the present PINT architecture and protocol, the service request to the PINT server is always initiated by the PINT client over the interface A. The PINT server can either be co-located with the IN service control or a similar entity, referred as executive system or communicate with it over the PSTN-specific interface E.

As Figure 3.33 shows, the PINT client and SPIRITS server are co-located in subscriber's IP host. In fact, both can be implemented to run as one process. No provision is made for interactions between the PINT client and SPIRITS server. Similarly, the PINT server and SPIRITS proxy are assumed to be co-located, too. This assumption is convenient but not essential; the PINT server could also be co-located with the SPIRITS client. In either case, no specific provision is made to define interworking between either the PINT server and SPIRITS proxy or PINT server and SPIRITS client other then by listing the overall PINT-related requirements [Fay00a].

3.7.3.1 The Example of PINT and SPIRITS Joined Work

The CCIB service allows users who are logged onto the Internet, using for example, a dial up account to determine the completion actions for telephony calls attempted to the telephone number that they are presently using for this connection. Examples of these completion actions might be:

- refuse the call,
- forward the call to voice mail,
- forward the call to another number,
- drop the Internet connection and take the telephony call over the same line,
- establish VoIP connection on terminating side to take the call,
- take details of the callee for later connection,
- take a voice message which can then be relayed to the terminating end device.

Both PINT and SPIRITS are intended to provide mechanisms for requests or initiation of service between the Internet and PSTN domains, and vice versa. Requests initiations from the Internet to the PSTN and responses to these requests is the PINT case. Requests from the PSTN to the Internet domain and responses to these requests is the SPIRITS case. This is shown more clearly in Figure 3.34.

3.7.3.2 General Requirements

The specification an initial set of requirements and service characteristics for an implementation of CCIB using the PINT and SPIRITS paradigms is given:

- Use of the service must be capable of being undertaken in a secure manner.
- Use of this service should be non repudiable.
- The described implementation should not be dependent in any way on specific technology being deployed outside of the Internet domain.

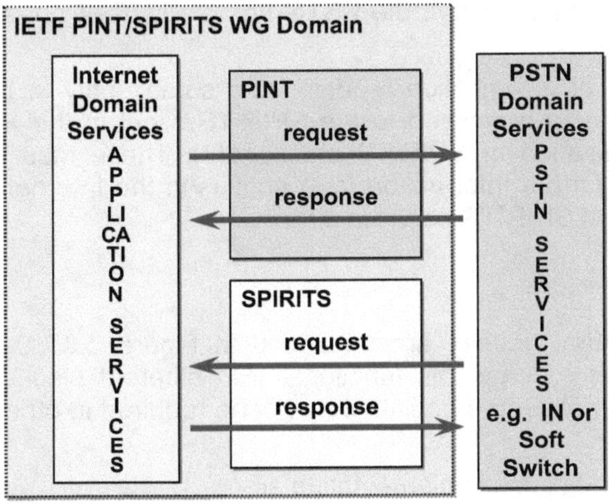

Figure 3.34 - PINT and SPIRITS bridging the PSTN/ Internet domains.

3.7.3.3 General architectural and service considerations and implications

This registration could be forwarded to the PSTN domain using the PINT protocol. Such a registration may be undertaken each time a subscriber connects to the Internet and wishes to receive SPIRITS like requests.

Possible implementations might be that the PSTN domain could dynamically set flags (service marks) in the switch relating to the subscribers line. This could indicate that call attempts to this number, for the duration of the current call will result in the initiation of a SPIRITS service.

An IN system could be used to handle call attempts to busy lines over which a subscriber is engaged in an Internet session.

It is entirely possible in some implementations, as a Soft-Switch providing Internet offload that no registration phase will be required.

In implementations where explicit subscriber registration is deemed to be required registration storage can be undertaken in several locations in either the Internet or PSTN domains. This data can either be: located in the Internet domain and referenced directly by a ICW/CCIB SPIRITS client also in the Internet domain once it has received a request from the PSTN, or located in the PSTN domain and referenced by some entity within this domain that then issues the request to the CCIB/ICW SPIRITS client in the Internet domain. In some implementations this request may require some form of resolution within the Internet domain to map the request to the current IP address of the subscriber.

The request initiated within the PSTN domain is itself not a SPIRITS request.

Upon receipt of an initiation request from some entity in the PSTN domain, the request is passed to the SPIRITS client in the SPIRITS gateway for formulation of SPIRITS message(s). These messages may be used to gain further information from entities in the Internet domain or to initiate/ request SPIRITS services directly.

3.7.3.4 Architecture

The architectural entities are described in Figure 3.35. Various implementations may locate this functionality in different places or domains, PSTN or Internet. Some entities may not be required in other possible implementations.

The entities within this implementation are:

- PSTN - An entity in the PSTN domain (e.g., IN or soft switch) initiates a request for service from this domain to the Internet domain. This does not mean that it directly passes SPIRITS requests. Instead requests are issued to the SPIRITS gateway where the SPIRITS client constructs and issues SPIRITS requests.
- User equipment - In this implementation the user connects to Internet using a dial up account from equipment suitable for such connectivity. User equipment will have a small SPIRITS server for receiving SPIRITS requests and possibly a PINT client for registration and initiation of PINT services. These entities may have been previously downloaded or may be downloaded each time a subscriber registers that they are connected and wish to be able to use PINT and SPIRITS services.
- Web server - Provides a mechanism for subscribers to register for receipt of SPIRITS messages.
- SPIRITS client - Contained within the SPIRITS gateway (Figure 3.33), it receives initiating requests from the PSTN and formulates them into SPIRITS requests.
- SPIRITS server - Receives and handles SPIRITS requests. As previously stated, this server could either be initially downloaded and used after each subsequent registration, or, downloaded each time as part of the registration process.
- Subscriber data store - Maintains subscriber details and registration requests.

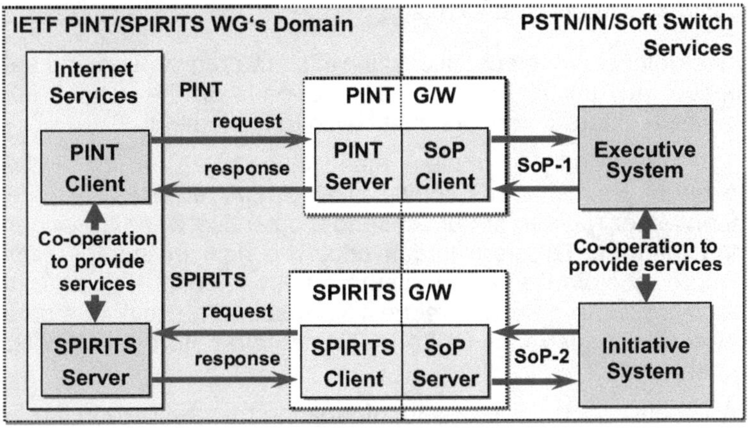

G/W – Gateway; SoPx – Some other Protocol

Figure 3.35 - Architectural differentiation between the PINT and SPIRITS protocols and PSTN/IN interface.

3.7.3.5 Implementation

Figures 3.36 and 3.37 briefly detail an implementation for provisioning the call completion Internet busy (CCIB) service. These figures identify the objects and the sequence of flows required and are split into two for clarity. The first provides details of how registration for use of the service is made. The second details how the user is contacted after a call attempt has been made, and how their choice, as to how to handle the call, is then relayed to the initiative system.

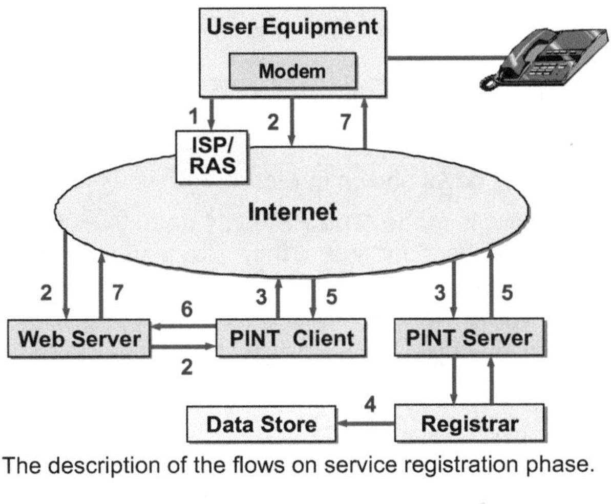

Figure 3.36 - The description of the flows on service registration phase.

3.7.3.6 Service Registration Phase

There are at least two mechanisms, in which the registration and setup for this service may occur. Each of these depends on the various business models which different vendors might wish to implement:

- The service may be implemented by use of an explicit binding to a telephony subscriber's known billable number. The customer may then either register or subscribe to the service when they connect to the Internet. This registration could be held in the PSTN domain where receipt of a call attempt to a busy telephone number may directly result in a service request, or result in a check of a data store of current registrands in the PSTN domain before the service is requested.
- Alternatively, the registration information may be held in the Internet domain, and the PSTN system may query this store, and activate the service, when the line is found to be engaged.
- This type of implementation is inherently more secure than the next option due to the explicit binding to a telephone number. Mechanisms can more easily be provided to require registrations from this line for the subscriber.
- The second scenario could allow a subscriber to register at whatever telephone number they happen to be using for their dial up connection. Once on-line they would register to the service.
- This registration could contain the currently used telephone number though this provides further security risks (e.g., wrong number specification by fault or design).
- This proposal could allow subscribers to register in a portable fashion from any number.

3.7.3.7 Call Attempt Phase

The description of the flows shown in Figure 3.37 is as follows:

- The call attempt is made. The PSTN/IN detects call in progress and busy SPIRITS subscriber line. It then plays an announcement to the calling party. Next, the PSTN/IN initiates a request to the SPIRITS client, using some other protocol, which could be INAP.
- The SPIRITS client issues the SPIRITS invitation request.
- Presently registered details are looked up.
- The final SPIRITS invitation request is constructed and sent. Then one of two things can occur detailed in 4a and 4b.

- The invitation request is received by the called party.
- The invitation request does not reach the intended recipient or times out. A failure message is returned to the SPIRITS client on the initiative system.
- The SPIRITS server applet on the called party's machine receives the invitation request and allows the user to decide how this call should be handled. Possible options here could be:
    - drop the Internet connection and take the call over the PSTN/IN system,
    - accept the connection via a VoIP gateway in the PSTN/ IN,
    - pass the call on to a voice mail system,
    - play announcements,
    - refuse the call.
- The called party's wishes are contained in this response which is passed back to the initiative system.

Due to the fact that registration is issued by the subscriber, not hand crafted or permanently pinned-up by the service provider, this inherently provides the possibility of service mobility. A subscriber may register themselves on any number [Bul00].

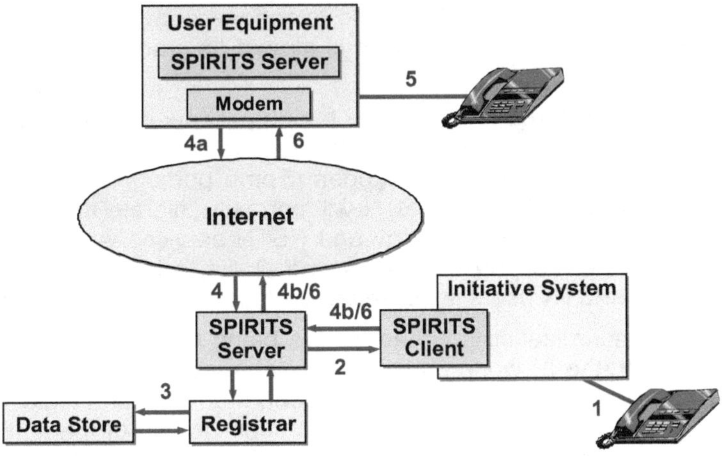

Figure 3.37 - The description of the flows on call attempt phase.

3.8 Converged Network Services Modeling

In last decade, the Internet has proven its ability to carry real-time data, including voice. Today, a small amount of voice traffic has already been diverted from PSTN to the Internet. If it expands, this phenomenon could completely change the rules of the game for telecommunications.

The Internet telephony is becoming a successful voice technology as evidenced by the burgeoning market for computer-based telephony products. This was enabled by recent advances in different technologies. In the signal processing field, new speech compression standards allow voice signals to be coded at very low bit rates while keeping their quality acceptable for conversational services. Moreover, the increasing bandwidth in IP access networks associated with the increasing routing capacity in the IP backbone makes it possible to reach an interactivity level similar to that offered by circuit switched networks. In addition, the dramatic growth of IP terminals with expanding processing power, memory, and multimedia capabilities allows IP-based voice services to be deployed on a very large scale.

On the other hand, the PSTN has made very impressive achievements in terms of coverage, reliability, and ease of use. The availability of the service is such that users are accustomed to receiving dial tone every time they pick up the phone and to being connected to any selected called party. PSTN terminals are also usable by most disabled people and people with limited education. In addition, the telephone network is being extended by cellular networks, this growth is almost as dramatic as that of the Internet.

Matching these features with a fully IP based network is a major engineering challenge. Meeting it may take next decade; in fact, there is no consensus today that this will ever happen. Some portion of the voice services currently offered by the PSTN will certainly migrate to an IP-based technology. However, IP telephony and PSTN services will coexist for a considerable time. For these reasons, the ability to interconnect IP telephony users to PSTN users is essential [Bri05].

Two main standardization approaches are being carried out for IP/PSTN interworking. In the IP world driven by the Internet engineering task force (IETF), interworking with the PSTN has been the result of a logical extension to the IP telephony service, which is seen as one of many IP applications. IPTEL, MMUSIC, PINT, and SPIRITS are the main IETF working groups concerned with IP telephony. In the telecommunications world, the international telecommunication union/telecommunication standardization sector (ITU-T) and the European telecommunications standards institute

(ETSI) are the main contributors in terms of standards and pre-standard documents.

The ITU-T has initiated various standardization activities that captured the attention of most of the industrials involved in the field. Related to these standards, the ETSI project telecommunications and Internet protocol harmonization over networks (TIPHON) undertook the effort to identify additional technical agreements required for the interoperability between IP networks and circuit switched networks. Some industrial consortia such as the international multimedia teleconferencing consortium (IMTC) through its voice-over-IP (VoIP) group also provide recommendations related to the implementation interoperability that is required in a multi-vendor context [ITU-H3, ITU-H22, ITU-H24, ITU-H23, ITU-H4].

3.8.1 Scenarios for Services Interworking in Converged Networks

Interworking of IP and IN voice services can be considered as part of a much bigger effort undertaken by standardization bodies in the field of network and service interworking. The most obvious interworking scenario between IP and the IN is when the PSTN connection is used as a lower data layer by the access part of an IP network (for example, dial-up access to an Internet service provider). In the context of IN and IP telephony services, interworking is the ability to offer a broader service that results from their peer juxtaposition. The hybrid voice services provide connectivity between users of both networks as well as between users of the same network given that part of the communication uses the service of the other network. Therefore, hybrid voice communications involve both PSTN and IP voice services and/or both types of terminals.

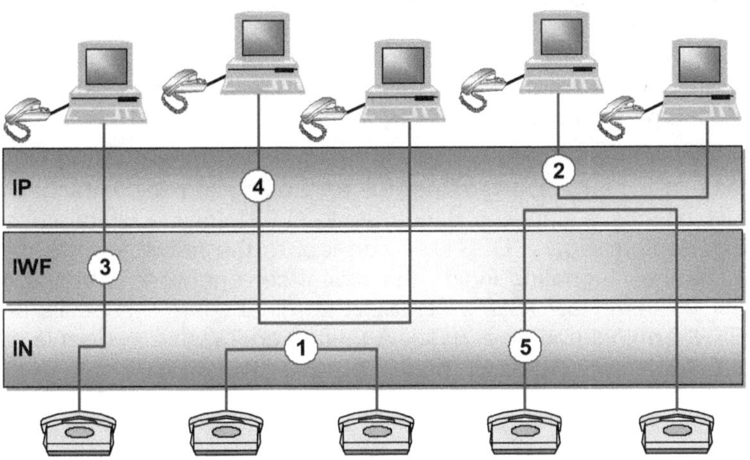

Figure 3.38 - Basic IN/IP communication scenarios.

Figure 3.38 illustrates five basic voice communication scenarios. Hybrid voice services are represented by scenarios 3, 4, and 5. In these scenarios, an interworking function (IWF) is needed to perform all protocol conversions and data adaptations. The IP and IN areas represent a protocol concept and do not necessarily involve a real network. Therefore, an IWF device may be used to connect two networks (i.e., a network adaptor) or a terminal to a network (i.e., a terminal adaptor).

For voice services, the IWF provides the following mechanisms:

- Signaling adaptation consists of the processing and translation of incoming signaling messages. It mainly concerns the call setup and clearing phases.
- Media control consists of identifying, processing, and translating service-specific control events that may be generated by the user or the terminal.
- Media adaptation consists of adapting the voice data to the data transfer channel of the downstream network [Kry07b, Ryu06].

3.8.1.1 Scenario 1

In scenario 1, two standard phone sets are connected via the IN/PSTN. The PSTN core network is based on a circuit switched network in which each circuit corresponds to a 64 kbps digital channel. An IN/PSTN terminal can be either digital or analog. Standard phone sets are attached to the PSTN by means of an analog access network, which merely corresponds to the set of subscriber loops (the copper wires that link the customers to the central office). On an analog access network, voice is transmitted as a 3 kHz wideband analog signal and is digitized at the access switch. In this case, signaling capabilities on the analog part of the access network (for example, address notification) are reduced to in-band coding of dual tone multi-frequency (DTMF) tones.

Also, ISDN allows voice terminals to have digital access to the PSTN. In this case, a digital voice terminal (or an analog terminal attached to an adaptor) initiates a signaling dialog using Q.931 (or the digital subscriber signaling system No.1 - DSS1) to connect to the network via a 64 kbps digital channel. Signaling inside the digital core network is based on the signaling system No.7 (SS7). An ISDN terminal seamlessly calls an analog PSTN terminal and vice versa. A unified addressing system is defined in ITU-T Recommendation E.164 [Kry03a, Kry04, Ive05].

3.8.1.2 Scenario 2

Scenario 2 illustrates what is generally referred to as IP telephony. IP telephony follows the IP paradigm: all service-specific processing and protocols, such as signaling and media coding, are pushed to the end systems and are transparent to the network. Applications can be built on top of the transmission control protocol (TCP) or the user datagram protocol (UDP), depending on whether they are loss sensitive or time-sensitive respectively.

For example, the TCP transport protocol is used to carry the signaling stream since the signaling channel has to be error-free. However, because of its intrinsic timing constraint, voice traffic is usually transmitted over UDP. The time-continuous property of voice signals requires that the transport channel ensure the appropriate streaming needed for data resynchronizing at the receiver. For this reason, the real-time protocol (RTP) is used. The sequence numbering field of RTP packet headers is used to reorder the receiving packets in case of out-of-sequence delivery, UDP does not ensure packet sequencing; the time-stamp field indicates the temporal playback position of the data payload. In addition, RTP allows the receiver to identify the media coding type (that is, which voice coding standard has been used at the coder side).

As far as end users are concerned, personal computers are the most common IP terminals. The processing and control parts of an IP telephony terminal are therefore usually implemented in software. However, a standard telephone set can also be connected to an IP telephony service by means of a network adaptor that provides a minimal set of the required protocols. This has the advantage of having the potential to reach a much larger number of users than PC holders.

ITU-T Recommendation H.323 and its related set of standards for packet-based multimedia communications in addition to the several related efforts carried out by the ETSI, IETF, and the IMTC, certainly constitute the most advanced framework that covers essential IP telephony issues [Kry06a, Fu05, Lul06].

3.8.1.3 Scenario 3

In scenario 3, the two terminals involved in the call use different protocol stacks to communicate with their access networks. The protocol conversions occur at the networks' boundaries. Two terminals, of different types in this case, communicate with each other to ensure an ad-hoc voice service to the end users. Scenario 3 requires both the mapping of media and media control channels and the mapping between signaling protocols [Ejz04, Kry06b, Agb07].

3.8.1.4 Scenarios 4 and 5

In scenarios 4 and 5, the same protocols are used at the interface of each terminal, but a different protocol is used between them. The protocol conversions in both directions take place (at least twice) at the boundaries of the traversed networks and the presence of another network in the middle should be transparent to end users. In these scenarios, both the mapping of media and media control channels and the mapping between signaling protocols are generally required. However, mapping between signaling protocols can be avoided in some configurations. In particular, when the IP network is used only as a backbone network (scenario 4); all IN/PSTN signaling information can be transferred transparently through the IP network [3Com00, Ham99, Dix96, Fow96, Pet96, ITU-E1, ITU-Q121, Kry07b].

3.8.2 Signaling and Control for Different Scenarios

If two different signaling protocols are used in the interconnected networks, then the IWF translates the signaling messages in such a way that the end-to-end call can be completed. In the H.323 gateway, Q.931 is used in both the IP network and IN/PSTN/ISDN access. However, the Q.931 signaling channel between an IP terminal and the gateway is terminated at the gateway (that is, Q.931 messages are processed in the gateway and not simply forwarded). A peer Q.931 channel is then used to support the call control on the PSTN side. This is mainly due to the fact that H.323 has defined a particular use of Q.931 messages, so that there is not necessarily a perfect correspondence with the ISDN use of Q.931. Figure 3.39 on shows the IWF protocol stacks in the control plane in scenario 3.

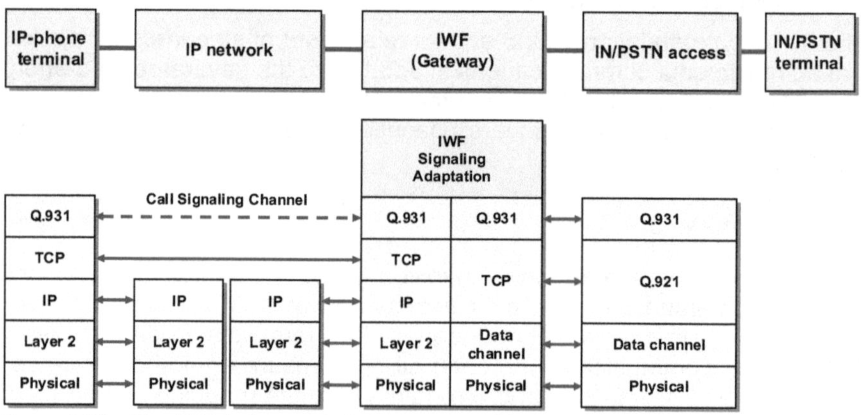

Figure 3.39 - The IWF in the control plane as an ISDN terminal to the PSTN.

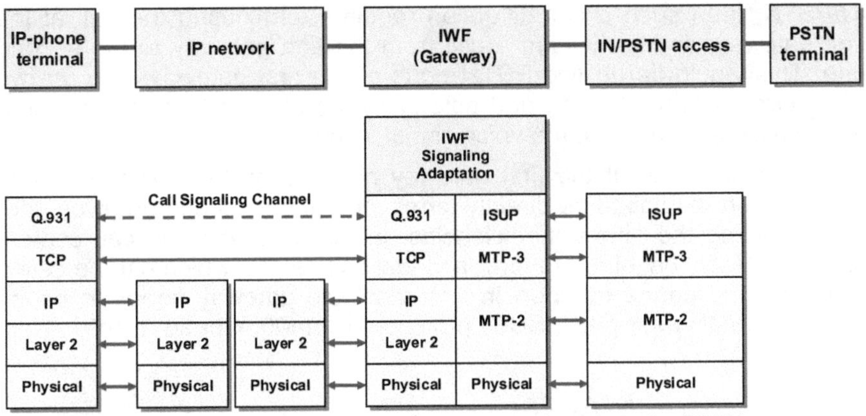

Figure 3.40 - The IWF in the control plane as a network node to the PSTN.

The IP/PSTN gateways are usually seen as administrative boundaries between a network provider, usually the operator, and a network custommer, usually a company or Internet service provider. For this reason, they are connected to the network as terminals. However, the gateway can be connected as a network node to the PSTN, to have access to its SS7.

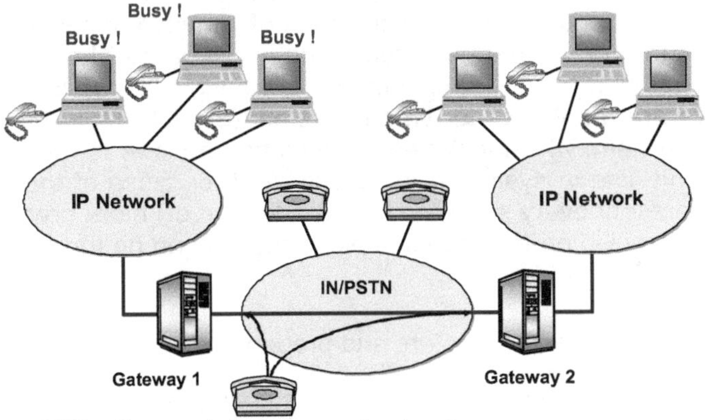

Figure 3.41 - SS7 call scenario in converged network.

Figure 3.40 shows the protocol stack needed for scenario 3 where the gateway connected to the PSTN is an ISDN node. For example, the scenario depicted in Figure 3.41, where two IP telephony-based call centers are shown; each is connected to the IN/PSTN through gateways. The two call centers are combined to form a single virtual distributed call center; if all the agents in one call center are busy, calls are to be diverted to the other one. If the gateways do not have access to the SS7 network of the

IN/PSTN, then such a call diversion requires terminating the call at the first gateway, and reinitiating a call from the first gateway to the second one. This would tie up two PSTN ports of the first gateway, use up two voice circuits in the PSTN, and potentially introduce a high delay due to the convoluted route that the voice signal follows.

The other example, if the first gateway has access to SS7, then it can simply divert the call to be directly terminated at the second gateway, thereby avoiding the above inefficiencies. In this way, the two call centers can seamlessly be joined to form a virtual call center, which can be called at a common phone number. In this case, the gateway needs to implement the ISDN user part (ISUP) protocol [3Com00, Gba98, Han97, Kos-98, Kry08, Mei05].

3.8.3 Modeling of the First Scenario

The first scenario is basic for intelligent networks services support. It is widely described in Section 2.9, using M/G/1/K/K and M/M/2/K/K systems.

3.8.4 Modeling of the Second Scenario Using M/G/1 System

A real-time service over IP network is illustrated by scenario 2 (in Figure 3.38) referred to as IP telephony. After the analysis of real-time traffic behavior over IP network, we choose the M/G/1 queueing system for the second scenario modeling. The coefficient of variation of service time distribution of the real-time traffic processing in different IP hosts differ from an exponential (e.g., as in simple M/M/1 queueing system). So, in order to have more precise values for time in system (e.g., delay) we need to apply M/G/1 system, which includes more comprehensive formulas for waiting time and time in system calculations. And, because of that, we have choice to present every separate host (node, server) more precise, its behaviour. Examples of that model implementation can be the SIP or H.323 server [Dav05, Ryu06].

3.8.4.1 M/G/1 system parameters and properties

The SIP/H.323 server can be assumed as a single server, with an infinite population of potential users, and users served in first-come-first-served order. The parameters to be specified are as follows:

- λ - average arrival rate of customers;
- $s_1, s_2, s_3, \ldots$- moments of service time about the origin.

There are used the following formulas for describing the service time distribution:

$$T_S = s_1$$
$$\sigma_{T_S} = \sqrt{s_2 - s_1^2} \qquad (3.1)$$
$$C_S^2 = \frac{s_2 - s_1^2}{s_1}$$

To check, if the system is stable ($\rho < 1$), it is necessary to calculate: $\rho = \lambda T_S$.

3.8.4.2 Mean and Variance of Waiting Time

The moments of waiting time (about zero) are denoted by w_1, w_2, w_3, ..., and their values can be calculated from the following recurrence relation:

$$w_k = \frac{\lambda}{1-\rho}\sum_{i=1}^{k}\binom{k}{i}\frac{s_{i+1}}{i+1}w_{k-i} \quad \text{where} \quad w_0 = 1 \qquad (3.2)$$

The first and second moments of waiting time are gotten from eq. (3.2):

$$w_1 = \frac{\lambda s_2}{2(1-\rho)} \qquad (3.3)$$

$$w_2 = \frac{\lambda}{(1-\rho)}\left(\frac{\lambda s_2^2}{2(1-\rho)} + \frac{s_3}{3}\right) \qquad (3.4)$$

It is necessary to notice, that to get the second moment of waiting time, we have to know up to the third moment of service time. In general, we need to know up to the *(i+1)th* moment of service time to calculate the *i*th moment of queueing or waiting time.

The mean and variance of waiting time can be calculated using the following formulas:

$$T_W = w_1 \qquad (3.5)$$

$$\sigma_{T_W} = \sqrt{w_2 - w_1^2} \qquad (3.6)$$

3.8.4.3 Average Time in System

The average time in system is the sum of the waiting time and the service time. The distribution of time in system is known as a convolution of two distributions, i.e. the distribution of waiting time and the distribution of service time. And, the moments of time in system about zero, which are denoted by q_1, q_2, q_3,..., can be calculated using another recurrence relation, i.e.:

$$q_k = \sum_{i=0}^{k} \binom{k}{i} w_{k-i} s_i \tag{3.7}$$

The first and second moments of time in system are as follows:

$$q_1 = w_1 + s_1 \tag{3.8}$$

$$q_2 = w_2 + 2w_1 s_1 + s_2 \tag{3.9}$$

From eq. (3.8) we obtain:

$$T = T_W + T_S \tag{3.10}$$

The variance of time in the system is [Tan95]:

$$\sigma_T = \sqrt{q_2 - q_1^2} \tag{3.11}$$

3.8.4.4 Server Busy Period

The M/GX/1 queue server goes through alternating periods of being idle and busy. An idle period starts with the departure of a job leaving an empty queue, and ends with the arrival of the next job; because of the memoryless property of the exponential distribution, the average length of an idle period is $1/\lambda$.

A busy period starts with the arrival of a job which finds an empty queue, and ends with the departure of the next job which leaves an empty queue. Its average length is denoted by g_1. More generally, an interval between an instant when there are j jobs present in the system and one of them is starting service, and the next departure which leaves an empty queue is called a busy period of order j. Let g_j be the latter's average length. Obviously, a busy period is a busy period of order 1.

An important property of any busy period of any order is that it does not depend on the scheduling policy. As long as the server is obligated to work whenever there are jobs present, and jobs are not allowed to leave the system before they are completed, the busy period is determined by the instants of arrival and the required service times. Exactly which job is being served at any moment is not important. The policy may also allow interruptions of service, provided that the later are eventually resumed without loss [Mit98].

The moments of the busy period length about zero are denoted by g_1, g_2, g_3,... (Figure 3.46). The average length of the busy period is given by

$$g_1 = \frac{s_1}{1-\rho} \equiv \frac{T_S}{1-\rho} \tag{3.12}$$

where, g_1 is average length of busy period.

It is necessary to notice that the average length of the busy period does not depend on the shape of the service time distribution; it just depends on T_S, the average service time. Looking now at the higher moments of busy period length, the particular service time distribution does have an effect. The second moment of the busy period is

$$g_2 = \frac{s_2}{(1-\rho)^3} \tag{3.13}$$

The denominator factor $(1 - \rho)^3$ shows that for high utilizations the length of the busy period is extremely variable. From g_1 and g_2 we get the variance of the busy period and its coefficient of variation:

$$\sigma_g = \sqrt{g_2 - g_1^2} = \sqrt{\frac{\sigma_{T_S}^2 + \rho T_S^2}{(1-\rho)^3}} \tag{3.14}$$

where, σ_g^2 is the standard deviation of busy period.

$$C_g^2 = \frac{\sigma_g^2}{g_1^2} = \frac{C_S^2 + \rho}{1-\rho} \tag{3.15}$$

where, C_g^2 is squared coefficient of variation of busy period. Equation 3.15 for C_g^2 gives some scale to the busy period variation.

The third (skewness) and fourth (kurtosis) moments of the busy period duration are as follows [Tan95]:

$$g_3 = \frac{s_3}{(1-\rho)^4} + \frac{3\lambda s_2^2}{(1-\rho)^5} \tag{3.16}$$

$$g_4 = \frac{s_4}{(1-\rho)^5} + \frac{10\lambda s_2 s_3}{(1-\rho)^6} + \frac{15\lambda^2 s_2^3}{(1-\rho)^7} \tag{3.17}$$

Skewness - a parameter that describes asymmetry in a random variable's probability distribution. The skewness of a random variable X is denoted as *skew*(X). It is defined as:

$$skew(X) = \frac{E\left[(X-\mu)^3\right]}{\sigma^3} \quad (3.18)$$

where μ and σ are the mean and standard deviation of X.

Kurtosis - the degree of peakedness of a distribution, defined as a normalized form of the fourth central moment of a distribution. It is a parameter that describes the shape of a random variable's probability distribution. The kurtosis of a random variable X is denoted as $kurt(X)$. It is defined as

$$kurt(X) = \frac{E\left[(X-\mu)^4\right]}{\sigma^4} \quad (3.19)$$

where μ and σ are the mean and standard deviation of X [Bro85].

3.8.4.5 Numerical Results

The numerical results for the expected time in system, which is sum of the waiting time in the queue and the service time, as well as standard deviations, and coefficients of variation, are given in Figures 3.42 ÷ 3.44.

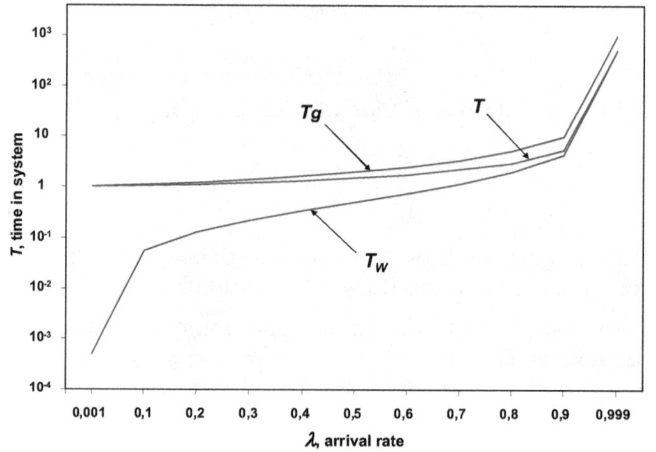

Figure 3.42 - The average waiting, time in system, and server busy period.

The variance, or standard deviation, and quantities related to it, are particularly important. Thus, a fundamental characteristic of a random variable is its variance. This is a measure of the spread, or dispersion, of the random variable around its mean. More precisely, the variance is defined as the average squared distance between random variable and its mean. Clearly, if random variable takes values close to the mean with high probability, then the variance is small (Figure 3.43).

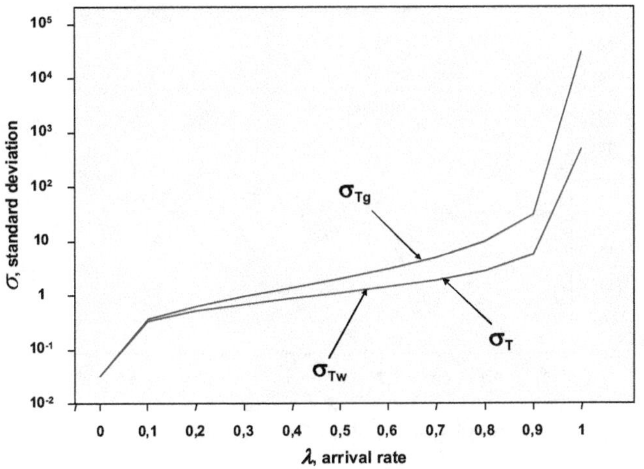

Figure 3.43 - The standard deviations of waiting, time in system, and server busy period.

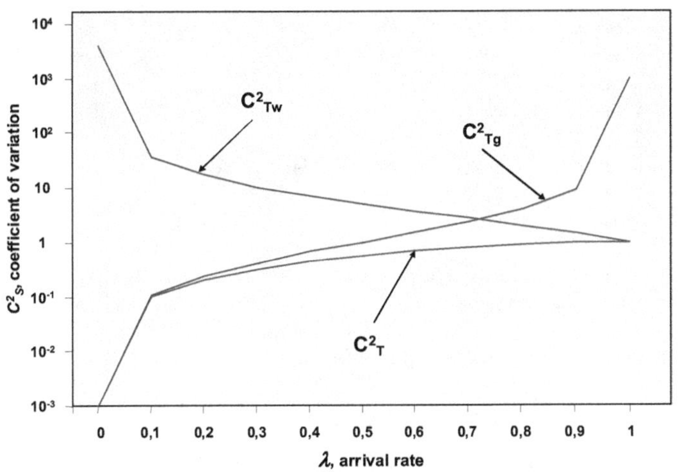

Figure 3.44 - Coefficients of variation of waiting, time in system, and server busy period.

It can be shown that, under certain quite general conditions, a random variable is completely characterized by the set of all its moments. In other words, if moment m_n is known for all n =1, 2,..., then the distribution function of random variable is determined uniquely (Figure 3.45) [Kry06a, Kry08].

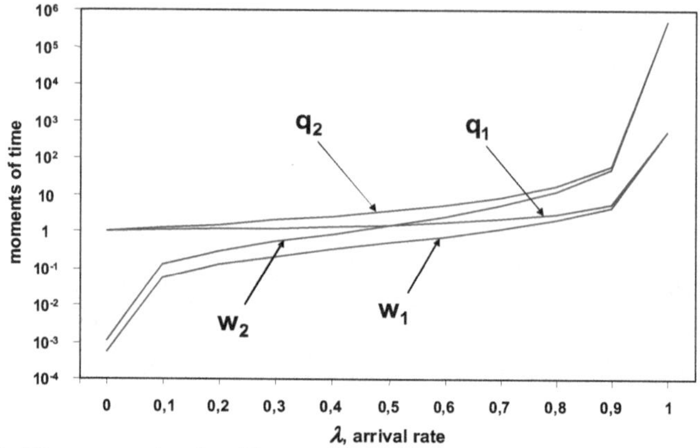

Figure 3.45 - The moments of waiting and time in system.

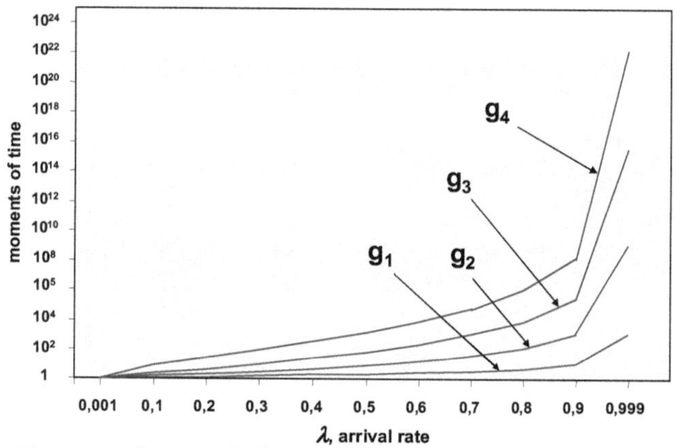

Figure 3.46 - The server busy period moments.

3.8.5 Modeling the Third Scenario Using M/E$_2$/1 system

We choose M/E_2/1 queueing system for the modeling of the third scenario because two different terminals are involved in the call, and, they use two different protocol stacks to communicate with their access networks. However, the service has to be considered as a single, composite. It means, from point of view of teletraffic theory that service is processed in two stages, and can be presented as an Erlangian queueing system [Dav06, Mei07].

3.8.5.1 Non Birth-Death Systems

Not all queuing models are birth-death systems. In some cases the Markov model will have transitions that are not nearest neighbor. However, they are still Markovian and have a steady-state solution given by $pq = 0$. Even though the model may not be birth-death, it still may have enough structure so that the steady-state equation will give rise to a small set of recurrence relations, although they will not be as simple as the birth-death ones.

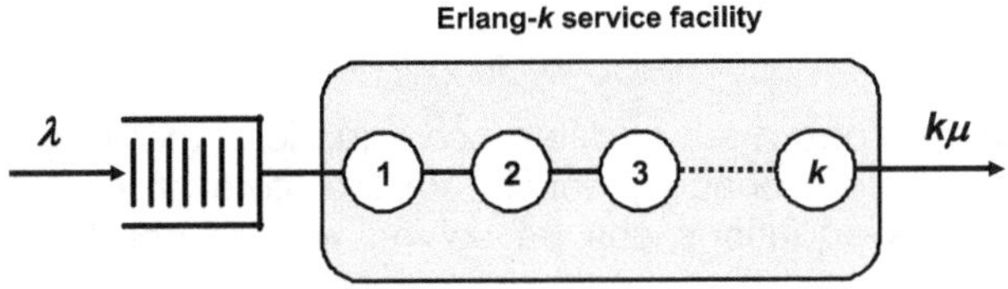

Figure 3.47 - The M/Ek/1 queueing system.

The problem still can be solved by using z-transforms techniques. In this case, to draw a transition diagram and write down the difference equations for the various states, corresponding to $pq = 0$.

In the $M/E_k/1$ model, customers arrive as a Poisson process, but are served by a single server, which takes a k-stage Erlangian distributed amount of time (Figure 3.47). Since the Erlang distribution is not memoryless for $k > 1$, the description of the state of the queue results in a non-Markovian process.

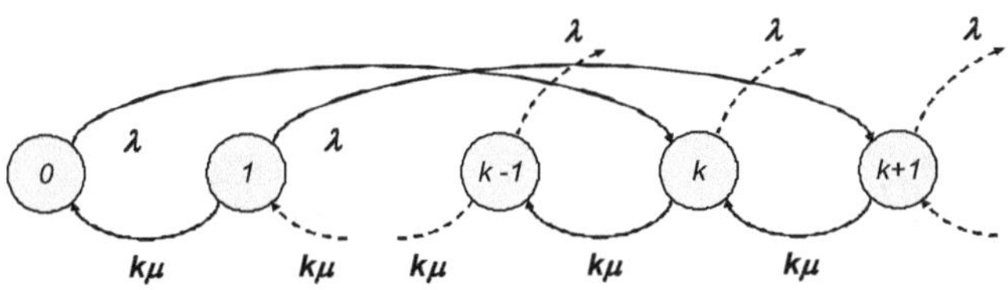

Figure 3.48 - The state transition diagram for the M/Ek/1 system.

The Erlangian distribution is the sum of identical and independent exponential distributions which are each memoryless. So if we change the concept of state, we can describe a Markov chain. The state is defined as the number of stages of service to be completed which are currently in the system. Under this interpretation each customer brings in k stages of service when it arrives. The state transition diagram is shown in Figure 3.48. The rate of decrease in the number of stages is always $k\mu$. Since no other customer may receive service until the current customer has completed all stages of service, precisely one stage of service is completed at a rate $k\mu$.

Therefore, completing all k stages takes $1/\mu$ on the average, maintaining the average total service time of $1/\mu$.

If it is no longer a birth-death system, we cannot simply apply the general birth-death equations. Instead, it is necessary to return to the equation $pq = 0$. There are two basic forms for the state equations from this balance equation.

$$\begin{aligned} \lambda p_0 &= k\mu p_1 & & \\ (\lambda + k\mu)p_i &= k\mu p_{i+1} & & 0 < i < k \\ (\lambda + k\mu)p_i &= k\mu p_{i+1} + \lambda p_{i-k} & & i \geq k \end{aligned} \tag{3.20}$$

These equations form a set of difference equations that must be satisfied by the steady-state probability density for the state of the M/E_k/1 queue. These difference equations can be solved with established z-transform techniques.

$$\begin{aligned} \sum_{i=1}^{k-1} (\lambda + k\mu)p_i z^i &= \sum_{i=1}^{k-1} k\mu p_{i+1} z^i \\ \sum_{i=k}^{\infty} (\lambda + k\mu)p_i z^i &= \sum_{i=k}^{\infty} k\mu p_{i+1} z^i + \sum_{i=k}^{\infty} \lambda p_{i-k} z^i \end{aligned} \tag{3.21}$$

The sum of these two equations produces a single equation that can be solved for the z-transform of the steady-state probability density.

$$(\lambda + k\mu)\sum_{i=1}^{\infty} p_i z^i = \sum_{i=1}^{\infty} k\mu p_{i+1} z^i + \sum_{i=k}^{\infty} \lambda p_{i-k} z^i$$

$$(\lambda + k\mu)\sum_{i=1}^{\infty} p_i z^i = \frac{k\mu}{z}\sum_{i=1}^{\infty} p_{i+1} z^{i+1} + \lambda z^k \sum_{i=k}^{\infty} \lambda p_{i-k} z^{i-k}$$

$$(\lambda + k\mu)\left(\prod(z) - p_0\right) = \frac{k\mu}{z}\left(\prod(z) - p_{1z} - p_0\right) + \lambda z^i \prod(z)$$

$$\prod(z)\left[\lambda + k\mu - \frac{k\mu}{z} - \lambda z^i\right] = (\lambda + k\mu)p_0 - k\mu p_1 - \frac{k\mu}{z} p_0$$

$$\prod(z) = \frac{(\lambda + k\mu)p_0 z - k\mu p_1 z - k\mu p_0}{(\lambda + k\mu)z - k\mu - \lambda z^{k+1}} \tag{3.22}$$

We consider only the stage of the server and ignore the state of the queue. If the system is busy, then the server is in one of k stages. Since the average time spent in each stage is the same and exponentially distributed, an user will find the server in any one of those stages with equal probability ρ/k, or idle with probability $1-\rho$ [Mol88].

3.8.5.2 M/E$_2$/1 Queueing System Features and Properties

In addition to specifying the number of customers in the system, it is necessary also specify the number of stages remaining in the service faci-lity for the customer in service, it results in representing each customer in the queue as possessing *k* stages of service yet to be completed for him. The total time that a customer spends in the service facility is the sum of *k* independent identically distributed random variables [Kle75].

Also, it is important to observe the coefficient of variation of service time. The Erlang-*k* distribution has a coefficient of variation squared ranging from $C_S^2 = 1$, when it is the exponential distribution, to $C_S^2 = 0$ when it is a constant distribution. The coefficient of variation is chosen by selecting the value of *k*, since

$$C_S^2 = \frac{1}{k} \tag{3.23}$$

The complete freedom in setting C_S^2 does not exist, since *k* must be an integer, but a useful selection of values can be obtained. The minimum value for *k* is 1, which gives the exponential distribution. When *k* increa-ses, the distribution becomes more symmetrical and also more tightly concentrated around the mean. For very large *k* the distribution is effecttively a constant. If C_S^2, the squared coefficient of variation of service time is given, then *k* should be chosen to be the largest value of *k* such that

$$k \leq \frac{1}{C_S^2} \tag{3.24}$$

The increasing effect of *k* is non-linear, so that the difference between *k* = 1 and *k* = 2 is about the same as between *k* = 2 and *k* = 5, between *k* = 5 and *k* = ∞. Unless *k* is quite small its exact value is not very important.

3.8.5.3 Waiting Time in M/E$_2$/1 System

The M/E$_k$/1 formula for the average waiting time is derived from the M/G/1 formula for the average waiting time, which with $C_S^2 = 1/k$ gives the following result:

$$T_W = \frac{\rho T_S}{2(1-\rho)}\left(1 + \frac{1}{k}\right) \tag{3.25}$$

The same actions are taken for the variance of waiting time, by putting the moments of the Erlang-*k* distribution into the appropriate M/G/1 formula, we get the following expression for the standard deviation of waiting time:

$$\sigma_{T_W} = \sqrt{\frac{T_S^2 \rho(k+1)}{12(1-\rho)^2 k^2}(4(k+2)-\rho(k+5))} \qquad (3.26)$$

3.8.5.4 Average Time in System for M/E$_2$/1

The formula for the average time in system for M/E$_2$/1 is also taken from the M/G/1 formula. If we substitute $C_S^2 = 1/k$ and rearrange slightly, then we get the result specific to M/E$_k$/1:

$$T = \frac{T_S}{1-\rho}\left(1-\frac{\rho}{2}\left(1-\frac{1}{k}\right)\right) \qquad (3.27)$$

Also, for the variance of the time in system, we know the moments of the Erlang-*k* distribution, and we can put these into the appropriate M/GX/1 results to obtain a formula for the variance, which is

$$\sigma_T = \sqrt{\frac{T_S^2}{(1-\rho)^2}\left[\left(1-\frac{\rho(4-\rho)}{6}\left(1-\frac{1}{k}\right)\right)\left(1+\frac{1}{k}\right)-\left(1-\frac{\rho}{2}\left(1-\frac{1}{k}\right)\right)^2\right]} \qquad (3.28)$$

3.8.5.5 Server Busy Period in M/E$_2$/1 System

The average length of the busy period is independent of the precise shape of the service time distribution and depends only on the average service time, i.e.

$$g_1 = \frac{T_S}{1-\rho} \qquad (3.29)$$

The variance (e.g., standard deviation) of the busy period does, however, depend on the shape of the service time distribution. For Erlang-*k* service the variance is given by

$$\sigma_g = \sqrt{\frac{T_S^2}{(1-\rho)^3}\left(\rho+\frac{1}{k}\right)} \qquad (3.30)$$

3.8.5.6 Numerical Results

The numerical results of waiting, time in system, and server busy period as well as standard deviations distributions are given in Figures 3.49 ÷ 3.50 [Tan95, All90].

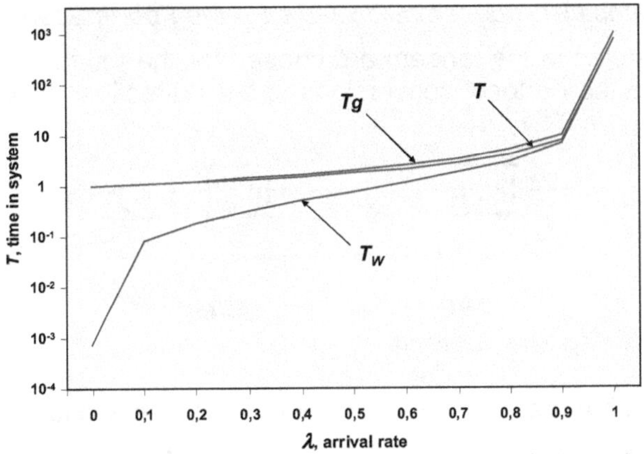

Figure 3.49 - The average waiting, time in system, and server busy period.

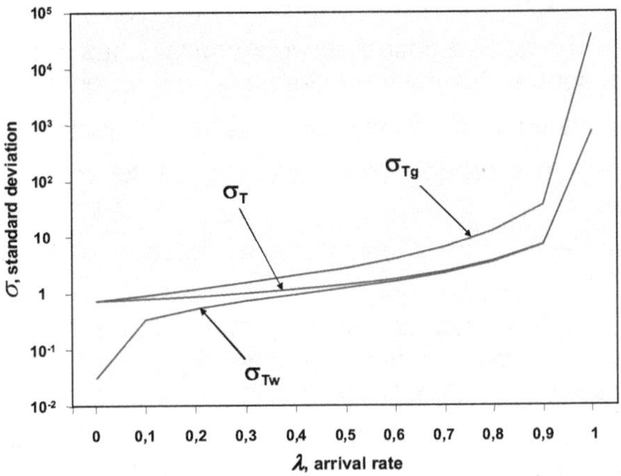

Figure 3.50 - Standard deviations of waiting, time in system, and server busy period.

Converged networks reduce costs by eliminating redundant hardware, communications facilities and support staffs. They also enable a new generation of integrated multimedia/voice/data applications. The modeling and analysis of system performance issues are useful tools in the development and engineering processes of these integrated applications. The analytical models have a high value in the early stages to uncover major performance problems which affect the design of the architecture before the cost of rectification is too high. The development of new modeling methods for rapid analysis is very important for the future wide deployment of converged networks applications and services [Kry07b, Noe07].

3.8.6 *Systems in Tandem M/G/1+ G/M/1 – the Fourth Scenario Model*

The two systems in the tandem are chosen for the fourth scenario modeling because the protocol conversions in both directions take place at least twice (Figure 3.51).

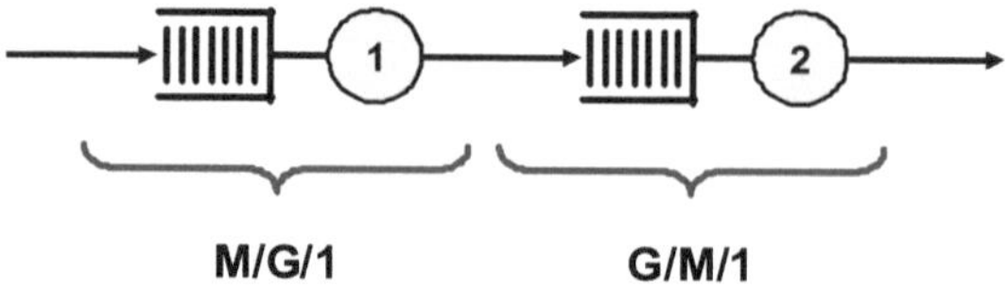

Figure 3.51 - Two systems in tandem.

The following features describe behavior two systems in tandem:

- When q_1 is M/M/1 or M/G/1 then q_2, is not. The arrival time (hence waiting time) is highly correlated, for example, a long packet is more likely to have a smaller waiting time at q_2;
- a queue in a packet-based network typically has packets from many different source-destination pairs passing through it;
- the packets arrive randomly from a variety of queues feeding into it;
- the depart in a random fashion to any of several queues this one feeds into;
- this also destroys the heavy correlations seen in the tandem queueing model;
- packets leaving a typical queue, q_1, can go to one of several queues next (e.g., q_2, q_3) or leave the network layer at this point (i.e. this node is the destination) [Vast].

The first system in tandem is M/G/1. It is described widely in Section 3.8.4. The second system is G/M/1. We describe the behavior and of that system in the next sub-chapters [Kry07a].

3.8.6.1 GI/G/1 System General Definitions

Till now, it was always assumed that the arrival process is a Poisson process. For other arrival processes it is seldom possible to find an exact expression for the mean waiting time except in the case where the holding times are exponentially distributed. In general it is assumed, that either the arrival process or the service process should be Markovian. For GI/G/1 queueing system it is in the state to give the theoretical boundary for the mean waiting time. By denoting the variance of the inter-arrival times as σ_a^2 and the variance of the holding time distribution as σ_h^2, the Kingman's inequality shows

$$T_W \leq \frac{aT_S}{2(1-a)}\left(\frac{\sigma_A^2+\sigma_S^2}{T_S^2}\right) \tag{3.31}$$

where, *a* is offered traffic. From that formula we can see that they are the stochastic variations, and that gives the reason for the waiting times. The formula (3.31) is the upper theoretical boundary. A good realistic estimation for the actual mean waiting time is obtained by Marchall's approximation:

$$T_W \approx \frac{aT_S}{2(1-a)}\left[\left(\frac{\sigma_A^2+\sigma_S^2}{T_S^2}\right)\left(\frac{T_S^2+\sigma_S^2}{T_A^2+\sigma_S^2}\right)\right] \tag{3.32}$$

where, T_a is the mean inter-arrival time ($T_a = T_{s/a}$). The approximation seems to be a downward scaling of Kingman's inequality so it just agrees with the Pollaczek-Khintchine's formula in the case M/G/1.

The example for a non-Poisson arrival process is the queueing system GI/M/1, where the distribution of the inter-arrival times is a general distribution given by the density function *f*(*t*) [Ive01].

3.8.6.2 GI/M/1 Queueing System State Probabilities

When we consider the system at an arbitrary point of time, the state probabilities will not be described by a Markov process only, because the probability that the occurrence of an arrival will depend on how long time has passed since the occurrence of the last arrival.

When the system is considered immediately before (or after) an inter-arrival time, there will be independence in the traffic process since the inter-arrival times are stochastic independent and the holding times are exponentially distributed. The inter-arrival times are balance points, and it is taken into consider the so-called embedded Markov chain.

The probability that immediate before an inter-arrival time to observe the system in state *i* is *p(i)*. In statistic equilibrium we will have the following result:

$$p(i) = (1-\alpha)\alpha^i \qquad i = 0,1,2,\ldots \tag{3.33}$$

where α is the positive real root, that satisfies the equation:

$$\alpha = \int_0^{\infty} e^{-\mu(1-\alpha)t} f(t)dt. \tag{3.34}$$

The steady state probabilities can be obtained by considering two for each of the following inter-arrival times t_1 and t_2. When the departure process is a Poisson process with the constant intensity *i*, with customers in the system, the probability $q(j)$ that there are j customers who have completed service between two inter-arrival times can be expressed by details in the Poisson process. Than, there can be set up the following state equations:

$$p_{t_2}(0) = \sum_{j=0}^{\infty} p_{t_1}(j) q(j+1),$$
$$p_{t_2}(1) = \sum_{j=0}^{\infty} p_{t_1}(j) q(j), \tag{3.35}$$
$$\vdots$$
$$p_{t_2}(i) = \sum_{j=0}^{\infty} p_{t_1}(j) q(j-i+1),$$

The normalization constant is as usual:

$$\sum_{i=0}^{\infty} p_{t_1}(i) = \sum_{j=0}^{\infty} p_{t_2}(i) = 1 \tag{3.36}$$

The $p(i)$ is not the probability to find the system in state i at an arbitrary point of time (e.g., time mean value), but to find the system in state i immediately before an arrival (e.g., call mean value) [Ive01].

3.8.6.3 Characteristics of G/M/1

The probability, that arriving customer finds the server busy θ is not the same as ρ, the server utilization for G/M/1, because of the general pattern of arrivals. Only the random arrivals have $\theta = \rho$. The value of θ can be obtained from formula.

$$\theta = f^*(s)\left(\frac{1-\theta}{T_S}\right) \qquad 0 \le \theta < 1 \tag{3.37}$$

where $f^*(s)$ is the Laplace-Stieltjes transform of the pdf of inter-arrival times. In some cases eq. (3.37) can be solved analytically, but in general a numerical procedure is required.

In Table 3.2 is presented probability that arriving customer finds the server busy for different distributions of inter-arrival time [Has80].

λ (μ = 1)	D/M/1 (C^2=0)	E_2/M/1 (C^2=0.5)	M/M/1 (C^2=1)	H_2/M/1 (C^2=2)	Ga/M/1 (C^2=5)
0.001	0.0000002	0.000021	0.001	0.0094	0.092
0.1	0.00005	0.00679	0.1	0.15728	0.526732
0.2	0.00698	0.04298	0.2	0.30757	0.626906
0.3	0.04088	0.10833	0.3	0.44786	0.698087
0.4	0.10736	0.19633	0.4	0.57442	0.755958
0.5	0.20319	0.30193	0.5	0.68377	0.805981
0.6	0.32424	0.42163	0.6	0.77442	0.850767

0.7	0.467	0.55291	0.7	0.84786	0.891784
0.8	0.62863	0.69394	0.8	0.90757	0.929949
0.9	0.8069	0.84334	0.9	0.95728	0.965877
0.999	0.974137	0.985459	0.999	0.99976	0.999947

Table 3.2 - The probability θ that arriving customer finds the server busy for different distributions of inter-arrival time.

3.8.6.4 G/M/1 Waiting Time - T_W

We can calculate the average waiting time using the following formula:

$$T_W = \frac{\theta T_S}{1-\theta} \tag{3.38}$$

The waiting time for the G/M/1 queueing system has a modified exponenttial distribution, so that

$$p(T_W < t) = 1 - \theta e^{-\frac{t}{T}} \tag{3.39}$$

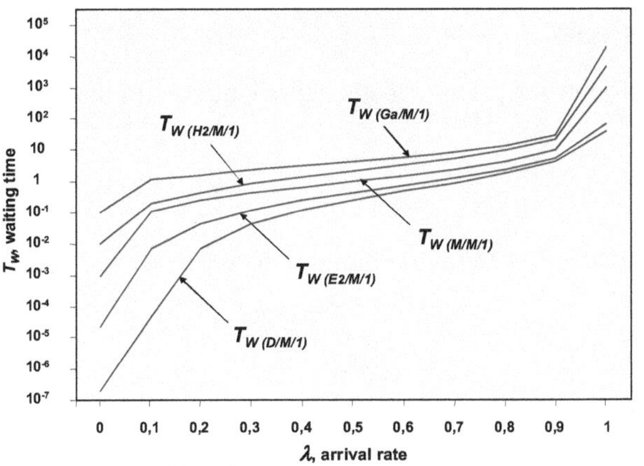

Figure 3.52 - The average waiting time.

And now, when we know the exact form of the distribution for waiting time, we can obtain standard deviation (e.g., variance) of time that user spends in queue (Figure 3.53):

$$\sigma_{T_W} = \sqrt{\frac{\theta(2-\theta)}{(1-\theta)^2} T_S^2} \tag{3.40}$$

It is necessary to notice, that the average waiting time increases when the arrival pattern becomes more irregular. Figure 3.52 shows average waiting time for different values of C_A^2. The effect of increased variance in the inter-arrival time is apparent, and is very marked at high utilizations.

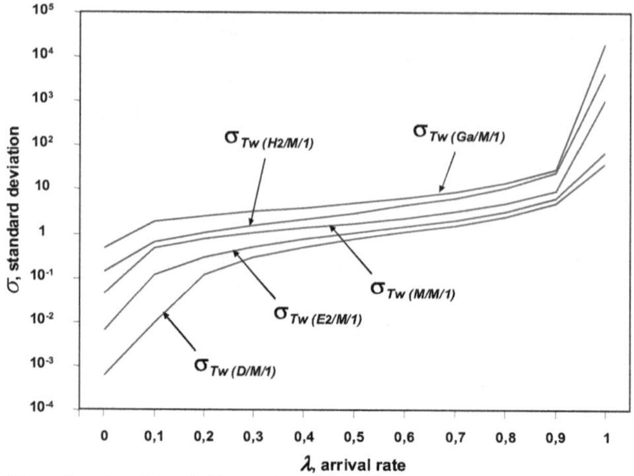

Figure 3.53 - The standard deviations of waiting time.

3.8.6.5G/M/1 Time in System – *T*

The average time in system for the G/M/1 queueing system has the following pattern (Figure 3.54):

$$T = \frac{T_S}{1-\theta} \tag{3.41}$$

in comparison with the M/M/1 system, where

$$T = \frac{T_S}{1-\rho} \tag{3.42}$$

we can see that for G/M/1 it is bigger:

$$\frac{T_S}{1-\theta} > \frac{T_S}{1-\rho} \tag{3.43}$$

because $\theta > \rho$, since θ is the probability that an arriving customer will have to wait.

The service time has an exponential distribution, so that

$$p(T < t) = 1 - e^{-\frac{t}{T}} \tag{3.44}$$

and, then the standard deviation (e.g., variance) time in system is (Figure 3.54):

$$\sigma_T = T \tag{3.45}$$

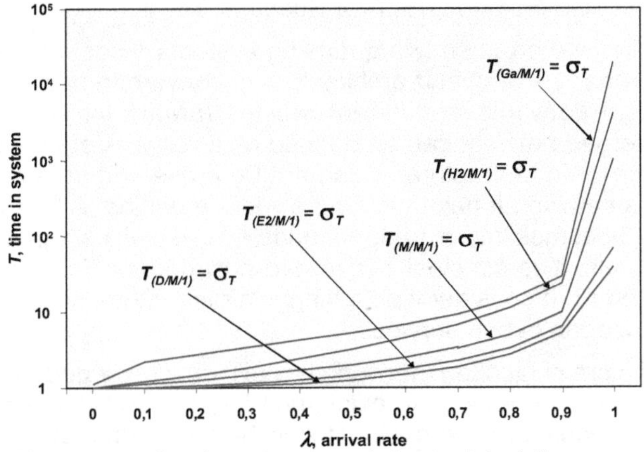

Figure 3.54 - The average time in system, equal standard deviation.

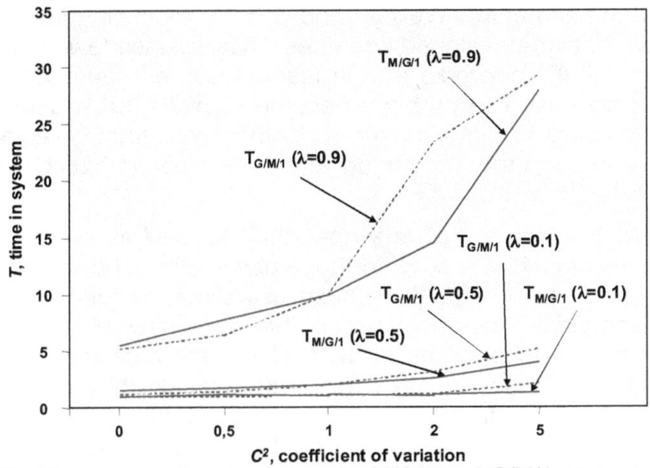

Figure 3.55 - The average time in system for M/G/1 and G/M/1 versus coefficient of variation.

Both the M/G/1 and G/M/1 systems show greater variability in longer times in system. The M/G/1 system demonstrates this for service times, while the G/M/1 model demonstrates this for inter-arrival times. Figure 3.55 plots average time in system against squared coefficient of variation. The M/G/1 curves are plotted against C_S^2, and the G/M/1 curves against C_A^2 [Tan95, Kle86a, Kry07a].

3.8.7 Systems $M/E_2/1 + H_2/E_2/1$ for Scenario 5

In this section, we propose two queueing systems model for performance design and analysis of hybrid architecture in converged network that span two intelligent networks in different regions (areas) through the IP network. Converged network can be defined as a network of networks. It has hybrid nature or infrastructure. It can not be presented as a single are or local access network. It has much more wider meaning. A "converged network" term becomes more understandable on services/applications and signaling levels. The simplest converged network consists from two networks: IN and IP. If consider it as a single infrastructure, so it is hybrid network and supports hybrid services.

The global communication network is built on converged (e.g., hybrid) networks. Where in reality can exist such hybrid networks? For example, two small regions still have circuit-switched infrastructures (e.g., INs). They are removed from each other. And, between them laying big, high-developed region, into which network operators has already deployed packet-based infrastructure. We can find a lot of examples like this nowadays, where in high-developed countries existing since few years pa-cket-based (e.g., all IP) networks and in less developed countries still PSTN only infrastructures. The problem become significant if a user from one small region wants to reach a user from other small remote region, he/she has to pass through that big region with packet-based backbone network [Kry06c].

In this case, the task of network operators, as well as service providers and traffic engineers is to provide these users with good QoS (e.g., minimum delay, but not only). Also, these users may require more than simple voice services. They are users of intelligent network, so they can require full range of IN services for both areas and between them. And, in order to support these remote users with services they expect, service providers and also traffic engineers have to take into consideration different aspects of network(s) performance. For this purpose, traffic engineer may require from service provider or service designer service scenario framework (e.g., signaling messages flow scenario and application/service execution program and so on), to analyze and predict additional loading for network/system.

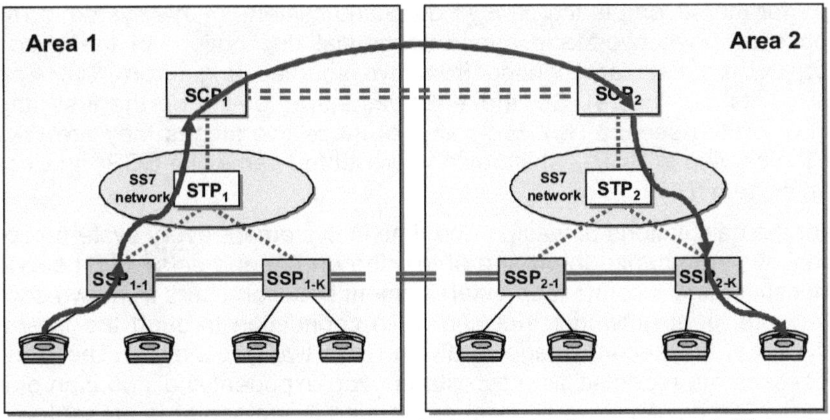

Figure 3.56 - Request for the connection establishment between users of different IN.

According to these requirements, we propose the service scenario framework (Figure 3.56), where the same protocols are used at the interface of each terminal, but a different protocol is used between them. The protocol conversions in both directions take place (at least twice) at the boundaries of the traversed networks and the presence of another network in the middle should be transparent to end users. In this scenario, both the mapping of media and media control channels and the mapping between signaling protocols are generally required. However, mapping between signaling protocols can be avoided in some configurations. In particular, when the core network is all-IP, all IN/PSTN signaling information can be transferred transparently through the IP network [3Com00, Ham99, Mee-06, Lio08].

For this scenario modeling we are applying the multidimensional birth-death model using method of stages. In case, when a user from the first network wants to get IN service from the second network, he/she has to go through two stages:

- first stage – call processing in the SCP_1;
- second stage – a request from SCP_1 has to be processed in SCP_2.

For the first stage we are applying the $M/E_2/1$ queueing system. The call processing in Intelligent Network is going through the several stages. And, in the typical SCP it goes through the two main stages: receiving and processing a request for IN service from SSP by SCP-server, and the processing of this request in SCP-database. However, the service (e.g., to be precise, a request for IN service) has to be considered as a single, composite. It means, from point of view of teletraffic theory, that service processing has to be presented as a two-stage Erlangian queueing system.

And, for the second is the $H_2/E_2/1$ queueing system. In this system arrival processes have two-stage hyperexponential distribution. In the second area IN, requests are coming from two sources (e.g., from SCP_1 and SSP_{1-K}, its own users). So, these requests are coming on the first stage (λ_2) or on the second (λ_1). From one of these two stages they are taken for the serving in SCP_2, which also is presented same like SCP_1 as Erlangian system (E_2).

After the calculations of waiting and time in system for every system separately, we summarize them and obtain the end-to-end delay of IN service processing and a connection establishment between users from two separated intelligent networks that can have connection through the Internet only. Also, the second stage is divided into two sub-stages. The arrival processes are modeled as a two-stage hyperexponential distribution of interarrival times. And, the request processing in the server is presented as two-stage Erlangian distribution.

The standard framework for two different IN, when the IP network is used as a backbone network, is shown in Figure 3.57.

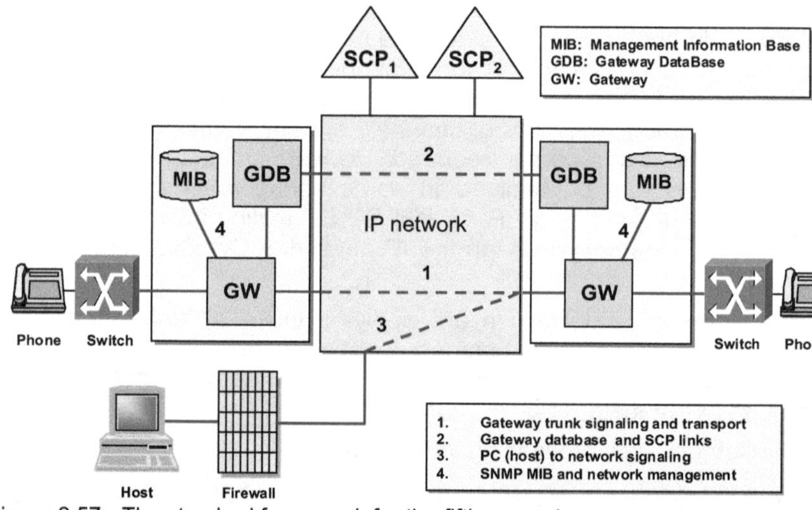

Figure 3.57 - The standard framework for the fifth scenario.

The main ideas behind this architecture are continuous data flow through the Internet, and interoperability among legacy telephone equipment and computers. Four groups of possible interactions were identified:

- gateway-to-gateway signaling and information transport,

- interactions between service control point (SCP) and gateway databases,
- host to network signaling,
- management information flow [Hub99, Ore07,].

3.8.7.1 Transformation of the Original Model into a Multidimensional Birth-Death Model Using the Method of Stages

The method of stages is an approximation procedure according to which a random variable with an arbitrary distribution is replaced by either a sum, or a mixture, or a combined sum-mixture of independent, but not necessary identical, random variables each being a phase or stage of the lifetime of the original random variable. This technique allows transformation of the original model into a multidimensional birth-death model, thus making it amenable to the powerful tools of analysis.

Lets consider, for example, the *s*-server Erlangian system; it is assumed that blocked customers are cleared, arrivals follow a Poisson process with rate λ, and service times are independent, identical, random variables with general distribution function. In this example of the application of the method of phases, it is assumed that the service time X can be approximated by a sum of n independent, but not necessarily identical, exponenttial random variables,

$$X = X_1 + X_2 + \ldots + X_n \tag{3.46}$$

and, the service time X is composed of n independent phases of service, the ith phase being exponentially distributed with distribution function

$$F_i(t) = P\{X_i \le t\} = 1 - e^{-\mu_i t} \tag{3.47}$$

Then

$$E(X) = \sum_{i=1}^{n} \mu_i^{-1} \quad \text{and} \quad V(X) = \sum_{i=1}^{n} \mu_i^{-2} \tag{3.48}$$

In particular, when $\mu_1 = \ldots = \mu_n$, then X has the Erlangian distribution of order n. Since it is true that

$$\left(\sum_{i=1}^{n} \mu_i^{-1}\right)^2 > \sum_{i=1}^{n} \left(\mu_i^{-1}\right)^2 \tag{3.49}$$

it follows that any service time described by a random variable X, where

$$E(X) > \sqrt{V(X)}, \quad (3.50)$$

can be represented as a sum of independent, exponential phases, as in (3.46), with given mean and variance. Furthermore, by judicious choice of the values n and μ_i (i = 1, 2,..., n), other moments might also be fitted to better approximate the given service time distribution. Of course, the phases $X_1,\ldots, X_n$ do not necessarily correspond to any actual phases of service, but are only artifices introduced for the purpose of approximating the original process by a birth-death process.

Now, let's suppose the service time X has greater variability than the exponential distribution prescribes; that is assumed in (3.50). In this case, the random variable X can be modeled as a parallel arrangement of exponential phases; that is, the realization of X is obtained by choosing, with probability p_i, the realization of the random variable X_i. Thus, the distribution function of X is

$$F_X(t) = \sum_{i=1}^{n} p_i F_i(t) \quad (3.51)$$

where, as before,

$$F_i(t) = 1 - e^{-\mu_i t} \quad (3.52)$$

then

$$E(X) = \sum_{i=1}^{n} p_i \mu_i^{-1}$$
$$\text{and} \quad (3.53)$$
$$V(X) = 2\sum_{i=1}^{n} p_i \mu_i^{-2} - \left(\sum_{i=1}^{n} p_i \mu_i^{-1}\right)^2$$

In this case, X is said to be a mixture of exponentials, and $F_X(t)$ is called the hyperexponential distribution function.

To continue with this example of the method of phases applied to the s-server Erlangian system, suppose that a representation of the form (3.46) has been fitted to the original data or hypothesized service time distribution. For ease of exposition let us assume n=2. Now, if we let $P(j_1, j_2)$ be the equilibrium probability that simultaneously there are j_1 customers in phase 1 of service and j_2 customers in phase 2, the corresponding conservation of flow equations are

$$\begin{gathered}(\lambda + j_1\mu_1 + j_2\mu_2)P(j_1, j_2) = \\ \lambda P(j_1 - 1, j_2) + (j_1 + 1)\mu_1 P(j_1 + 1, j_2 - 1) + \\ (j_2 + 1)\mu_2 P(j_1, j_2 + 1) \\ (j_1 + j_2) < s\end{gathered} \tag{3.54}$$

and

$$\begin{gathered}(j_1\mu_1 + j_2\mu_2)P(j_1, j_2) = \\ \lambda P(j_1 - 1, j_2) + (j_1 + 1)\mu_1 P(j_1 + 1, j_2 - 1) \\ (j_1 + j_2) = s\end{gathered} \tag{3.55}$$

Equations (3.54) and (3.55), together with the normalization equation,

$$\sum P(j_1, j_2) = 1 \tag{3.56}$$

can now be solved, numerically or otherwise, for the probabilities $P(j_1, j_2)$ $[(j_1 + j_2) \leq s]$, from which we can calculate the equilibrium probability P_j that there are j customers in service:

$$P_j = \sum_{j_1 + j_2 = j} P(j_1, j_2) = \sum_{k=0}^{j} P(k, j - k) \qquad (j = 0, 1, \ldots, s) \tag{3.57}$$

A particularly interesting aspect of the method of phases is that it can sometimes be used to obtain exact results by use of a limiting process. This idea has been developed by Schassberger (1973), who proved a theorem that states - any nonnegative random variable X can be represented as accurately as desired by a compound sum of independent, identical, exponential variables. The illustration of this idea by using the ordinary method of phases is done below to prove the important theorem (e.g., the Erlangian distribution),

$$P_j = \frac{\dfrac{(\lambda/\mu)^j}{j!}}{\displaystyle\sum_{k=0}^{s} \frac{(\lambda/\mu)^k}{k!}} \qquad (j = 0, 1, \ldots, s), \tag{3.58}$$

is valid for any service time distribution function with finite mean μ^{-1}.

And, next, the equations (3.54) and (3.55), which describe an Erlangian system in which service times are composed of a sum of two exponential phases with means μ^{-1} and μ^{-2}, are satisfied by the product solution

$$P(j_1, j_2) = \frac{(\lambda / \mu_1)^{j_1}}{j_1!} \frac{(\lambda / \mu_2)^{j_2}}{j_2!} c \qquad (j_1 + j_2) \leq s. \tag{3.59}$$

Substitution of (3.59) into (3.57) gives, after application of the binomial theorem,

$$P(j_1, j_2) = \frac{\left(\frac{\lambda}{\mu_1} + \frac{\lambda}{\mu_2} \right)^j}{j!} c \qquad (j = 0, 1, ..., s); \tag{3.60}$$

if we assume

$$\frac{1}{\mu} = \frac{1}{\mu_1} + \frac{1}{\mu_2} \tag{3.61}$$

will put it in (3.60) and normalize, the result is (3.58), as promised. And, finally, the assumption that the service time is composed of $n = 2$ phases is irrelevant; the same result (3.58) would be obtained for any number n of phases and corresponding mean values μ_i^{-1} whose sum is μ^{-1} [Coo90].

3.8.7.2 $H_2/E_2/1$

It is taken into consideration a queueing system that has the hyperexponential interarrival times. Using the method of stages, we assume there always exist customers available to enter our arrival mechanism, which consists of two arrival branches. Arrival branch 1 is chosen with probability q and has parameter λ_1; arrival branch 2 is chosen with probability $(1-q)$ and has parameter λ_2. Exactly one customer is in the arrival mechanism at all times. The system is represented graphically in Figure 3.58. To facilitate the analysis, let

- p_{njk} – probability of n customers in the system when the arriving customer is in arrival branch j and the customer in service is in service stage k, $n = 0, 1$; $j = 1, 2$; $k = 1, 2$;
- p_{0j} – probability of no customers in the system when the arriving customer is in arrival branch j, $j = 1, 2$.

The steady-state equations for this queueing system are written as [Whi-75, Kle86b]:

$$\lambda_1 p_{01} = 2\mu p_{112}, \quad \lambda_2 p_{02} = 2\mu p_{122}$$
$$(\lambda_1 + 2\mu)p_{111} = \alpha\lambda_1 p_{01} + \alpha\lambda_2 p_{02} + \alpha\lambda_1 p_{111} + \alpha\lambda_2 p_{121}$$
$$(\lambda_2 + 2\mu)p_{121} = (1-\alpha)\lambda_1 p_{01} + 1-(\alpha)\lambda_2 p_{02} + (1-\alpha)\lambda_1 p_{111} + (1-\alpha)\lambda_2 p_{121} \quad (3.62)$$
$$(\lambda_1 + 2\mu)p_{112} = 2\mu p_{01} + \alpha\lambda_1 p_{112} + \alpha\lambda_1 p_{111} + \alpha\lambda_2 p_{122}$$
$$(\lambda_2 + 2\mu)p_{122} = 2\mu p_{121} + (1-\alpha)\lambda_1 p_{112} + (1-\alpha)\lambda_2 p_{122}$$

And,

$$p_{01} + p_{02} + p_{111} + p_{121} + p_{112} + p_{122} = 1 \quad (3.63)$$

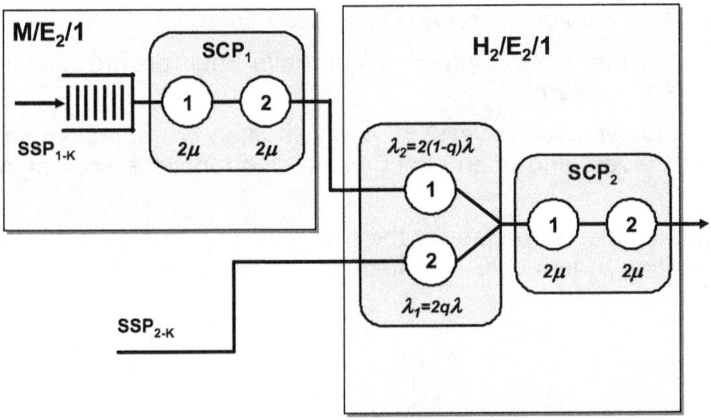

Figure 3.58 - Queueing systems model for the fifth scenario.

For the $H_2/E_2/1$ queueing system, there is a single server with FIFO queuing discipline. The number of requests for IN services is unlimited. Requests arrive at an average rate λ, so that the average inter-arrival time is $T_A = 1/\lambda$. The distribution of inter-arrival time is hyperexponential, with variance $\sigma^2_{T_A}$, and coefficient of variation squared for inter-arrival time is C_A^2. Service times have an Erlangian distribution, with average value T_S and coefficient of variation squared of C_S^2 [Tan95, Ryu08].

3.8.7.3 GI/G/1 Approximate Formulae of Waiting Time for $H_2/E_2/1$ Queueing System

In computer communications very often queueing problems may be represented by queueing systems of the type GI/G/1 (general input and general service process, single server). The mean waiting time and the probability of waiting are of the particular interest for system analysis or design.

The solutions for different traffic assumptions require the numerical evaluation of roots of transcendental equations by the aid of computers. For other types of arrival or service processes imlicit solutions are known (e.g., based on Lindley's integral equation). But often these solutions are not straightforward and require a lot of evaluation work. Therefore, the urgent need was felt to support the network planners and traffic engineer with simple explicit but general approximation formulae for the mean waiting time and the probability of waiting.

The restriction to the first two moments of the interarrival and service time distribution functions is because:

- for Poisson input the mean waiting time only depends on the first two moments of the service time distribution;
- the two-moments approximations have been proved useful e.g. for overflow systems;
- the models used for system analysis often are approximate models themselves, and often exact distribution functions are not known at all.

The approximation formulae for the mean waiting time T_W and the probability of waiting p_w in a GI/G/1 system are as following [Krä76]:

$$T_W = \frac{\rho T_S}{2(1-\rho)}\left(C_A^2 + C_S^2\right)\begin{cases} e^{-\frac{2(1-\rho)}{3\rho}\frac{\left(1-C_A^2\right)^2}{C_A^2+C_S^2}} & C_A^2 \le 1 \\ e^{-(1-\rho)\frac{C_A^2-1}{C_A^2+4C_S^2}} & C_A^2 \ge 1 \end{cases} \tag{3.64}$$

$$p_W = \rho + \left(C_A^2 - 1\right)\rho(1-\rho)\begin{cases} \dfrac{1 + C_A^2 + \rho C_S^2}{1 + \rho\left(C_S^2 - 1\right) + \rho^2\left(4C_A^2 + C_S^2\right)} & C_A^2 \le 1 \\ \dfrac{4\rho}{C_A^2 + \rho^2\left(4C_A^2 + C_S^2\right)} & C_A^2 \ge 1 \end{cases} \tag{3.65}$$

Under the heavy load conditions, waiting time has an approximately exponential distribution, with mean waiting time given by

$$T_W \approx \frac{C_A^2 + \lambda^2 C_S^2 T_S^2}{2\lambda(1-\rho)} \tag{3.66}$$

This approximation improves as ρ gets closer and closer to 1. For low utilizations the approximation is extremely poor.

The Marchal modification to the heavy-traffic approximation gives a formula:

$$T_W \approx \frac{\left(1+C_S^2\right)}{\left(1/\rho^2\right)+C_S^2} \frac{C_A^2+\lambda^2 C_S^2 T_S^2}{2\lambda(1-\rho)} \tag{3.67}$$

The Marchal, also Kleinrock vol.II - 1976, lower bound on average waiting time approximate formula is:

$$T_W \geq \frac{\rho^2 C_S^2+\rho(\rho-2)}{2\lambda(1-\rho)} \tag{3.68}$$

This bound will be non-negative only for service time coefficients of variation that satisfy $C_S^2 \geq (2-\rho)/\rho$. In our case (e.g., $H_2/E_2/1$ system, when $C_S^2=0.5$), it will always take negative values [Kle86b].

The formula given below is actually a special case of the Allen-Cunneen formula for G/G/m.

$$T_W \approx \frac{\rho T_S}{1-\rho}\left(\frac{C_A^2+C_S^2}{2}\right) \tag{3.69}$$

Next, we present numerical results we have got from mathematical analysis. We give here strictly theoretical results to see and to have good theoretical model of hybrid architecture in order to build stencil diagrams for the further practical usage. To be precise, we give normalized values of main parameters (e.g., arrival rates, service times, time in system). And, when applying real values of λ and μ, we have possibility to compare theoretical and practical results (for example, comparison of the theoretical values of time in system T, with practical ones), which gives great opportunity to correct and adjust real system parameters for the good performance of whole network.

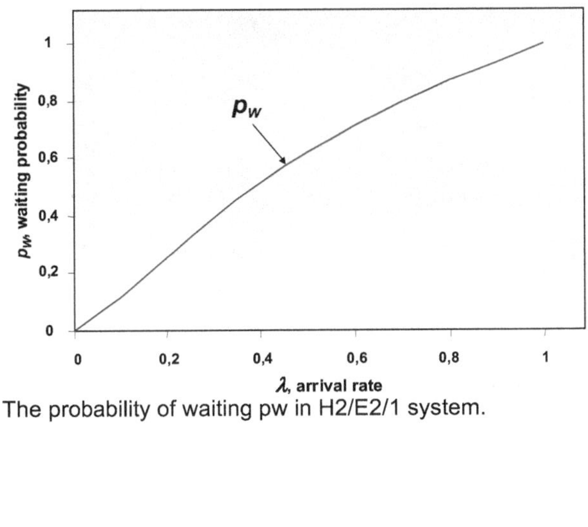

Figure 3.59 - The probability of waiting pw in H2/E2/1 system.

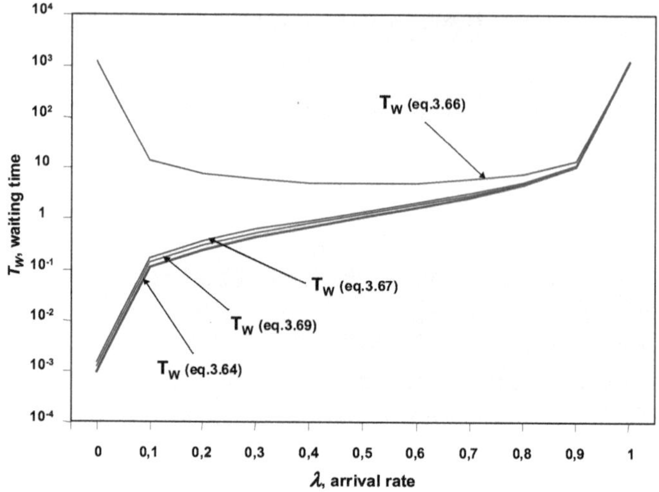

Figure 3.60 - The average waiting time in H2/E2/1 system under different traffic conditions.

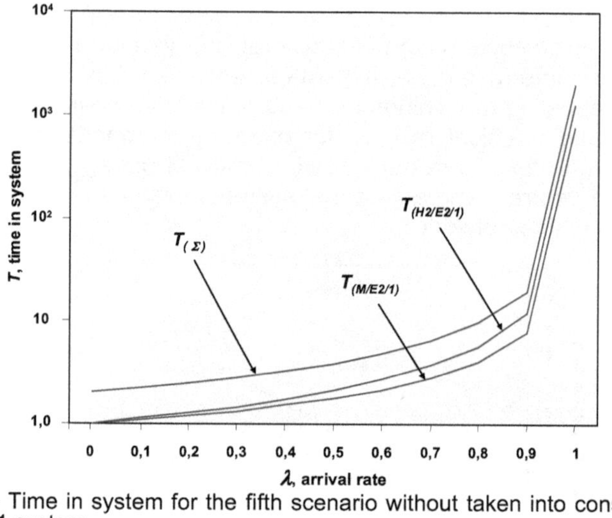

Figure 3.61 - Time in system for the fifth scenario without taken into consideration of pw in H2/E2/1 system.

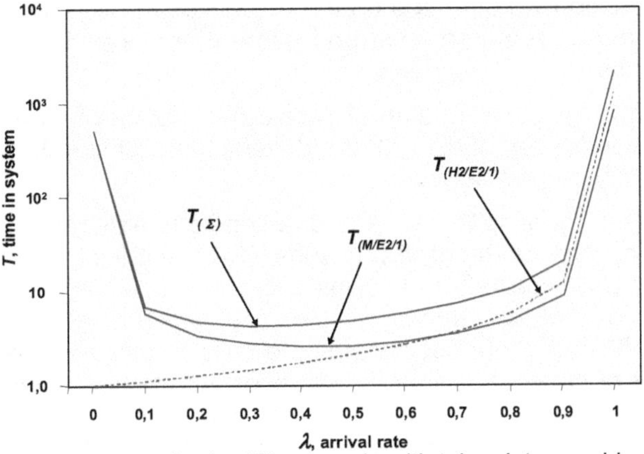

Figure 3.62 - Time in system for the fifth scenario with taken into consideration of pw in H2/E2/1 system.

In Figures 3.60÷3.62 we are giving comparison of the summarized waiting time and time in system (e.g., in M/E2/1 + H2/E2/1) without and with taken into consideration waiting probability p_w in H2/E2/1 system. According to eq. 3.65, we calculate waiting probability p_w for $C_A^2 \geq 1$ (e.g. $C_A^2 = 2$ for H2/E2/1 system). And, from Figure 3.59 we can see that p_w is growing with the increase of the server utilization, especially for arrivals from M/E2/1 system ($\lambda_2 = 2(1\text{-}q)\lambda$). So, for the second case (e.g., with taken into consideration p_w in H2/E2/1 system) we are doing backward calculations of M/E2/1 time in system. We have λ and p_w or q for H2/E2/1, and, from that, we can calculate λ_2, which brings us to $T_{M/E2/1}$ ($T_{M/E2/1} = 1/\lambda_2$). Finally, from Figure 3.62, we can see that $T_{M/E2/1}$ and $T_{(\Sigma)}$ have form as under the heavy load conditions. It is looking logical because requests for call processing in SCP_2 are coming from two sources, e.g. from its own SSPs and from SCP_1. Of course, an arrival rate from $SCP_{2\text{-}K}$ (λ_1) will be higher than λ_2, so requests from SCP1 will have to wait for processing longer time. And, $T_{(\Sigma)}$, because of that, is under the heavy load conditions [Oec07, Bhu07, Bou08].

3.9 Assessment and Modeling Results Summarizing

In this chapter, convergence between public switched telephone network (PSTN), its applications/services platform IN, and packet-based IP networks has been presented. Different standardization and research efforts of IN/IP integration have been described in details.

In the introduction we stressed on the objectives, enablers, network configuration, features requirements, and some other open issues of convergence process.

In Section 3.2 we gave a comprehensive description of work of the international and industrial bodies for the development of standards and architectures for converged networks.

Call control or signaling in converged networks is a very specific. In Internet telephony, the call control functions of a traditional circuit switch are replaced by a device referred to, as a call agent, a SIP server, a H.323 gatekeeper, or a feature server. This device, which is referred to as a call agent is an IP entity that coordinates the calls. A call agent executes a finite number of state transitions as it processes the call. These state transitions constitute its call model. The term call model when applied to an Internet call agent is a misnomer; a better term would be a protocol state machine. Unlike a traditional switch armed with an IN call model, the protocol state machine on a call agent does not contain IN specific trig-gers and states. Also, the number of call-related states of an Internet call agent is much less than those of the IN call model. Section 3.3 presented different call models and interworking between them for transparent service access from various underlying networks.

In Section 3.4 we described service control functionalities in terms of componentware model (e.g., service platform). The service control consists of two components: call control and messaging control. The call control offers a common call control interface to all services whereas the message control component provides an abstract interface to services in the service control layer.

Java is the basis for secure opening of IN to data networking. The JAIN abstracts the protocols covered by the protocol APIs into a single call control, coordination, and transaction model to be used by compliant services. In Section 3.5 we presented JAIN APIs as an attempt to abstract the network-level functionalities from the idiosyncrasies of different protocol variants so the changes in the underlying networks are transparent to the application.

Section 3.6 gave an overall description of services that can be recognized as converged. These services are divided into two groups: IN service requests towards IP networks, and IN service requests from IP networks. According to that division, we explained in detail signaling messages flows for every service scenario.

In the Section 3.7, two protocols PINT (for IN service requests from IP networks) and SPIRITS (for IN service requests towards IP networks) have been presented. These protocols are created for converged services support. They are working on the application layer of OSI model (e.g., above SIP protocol). PINT and SPIRITS protocols are mirror images of the integration process. And, their development is moving towards single, joined protocol in converged environment. An example of PINT/SPIRITS joined work is given in this section.

In the next Section 3.8, we analyzed, evaluated, and applied teletraffic for converged network services interworking. In the context of IN and IP services, interworking is the ability to offer a broader service that results from their peer similarities comparison. The hybrid services provide connectivity between users of both networks as well as between users of the same network given that part of the communication uses the service of the other network. Therefore, hybrid communications involve both IN and IP services and/or both types of terminals.

Thus, we have presented and illustrated in detail five basic scenarios for services interworking in converged network. We have also analyzed and evaluated signaling protocols for different scenarios.

Sections 3.8.3 – 3.8.7 dealt with different queueing systems we apply to differrrent service scenarios. We analyzed, evaluated and motivated usage of every queueing system into every scenario implementation. Also, we provided numerical results of important parameters calculations (e.g., time in system, waiting time, server busy period, number of users in system, number of users waiting in queue, and so on). And, we demonstrated the calculations results in numerous diagrams.

The first scenario is basic for intelligent networks services support. It was widely described in Chapter 2, using M/G/1/K/K and M/M/2/K/K queueing systems. So, Section 3.8.3 gave only brief information about that.

The second scenario (Section 3.8.4) is basic for voice over IP network. After the analysis of real-time traffic behavior over IP network, we have chosen the M/GX/1 queueing system for this scenario modeling. The coefficient of variation of service time distribution of the real-time traffic processing in different IP hosts differ from an exponential (e.g., as in simple M/M/1 queueing system). So, in order to have more precise values for time in system (e.g., delay) we applied M/GX/1 system, which includes more comprehensive formulas for waiting time and time in system calculations. And, because of that, we had choice to present every separate host (node, server) more precise, its behaviour. Examples of that model implementation can be the SIP or H.323 server.

For the third scenario modelling (Section 3.8.5) we used $M/E_2/1$ system, which is the special case of M/G/1 queueing system. Two different terminals are involved in the call, and, they use two different protocol stacks to communicate with their access networks. However, the service has to be considered as a single, composite. It means, from point of view of teletraffic theory that service is processed in two stages, and can be presented as an Erlangian queueing system.

The two systems in the tandem were chosen for the fourth scenario modeling because the protocol conversions in both directions take place at least twice. Both the M/G/1 and G/M/1 systems show greater variability in longer times in system. The M/G/1 system demonstrates this for service times, while the G/M/1 model demonstrates this for inter-arrival times.

For the fifth scenario modeling we applied the multidimensional birth-death model using method of stages. The fifth scenario presented two separated intelligent networks that have connections through packet-based network, e.g., the Internet. In case, when an user from the first network wants to get IN service from the second network, he/she has to go through two stages: the first stage – call processing in the SCP_1; the second stage – a request from SCP_1 has to be processed in SCP_2. For the first stage we are applying $M/E_2/1$ queueing system. And, for the second – $H_2/E_2/1$ queueing system. After the calculations of waiting and time in system for every system separately, we summarized them and got end-to-end delay of IN service processing and a connection establishment between users from two separated intelligent networks, that can have connection through the Internet only. Also, the second stage was divided on two sub-stages. The arrival processes were modeled as a 2 stages hyperexponential distribution of interarrival times. And, the request processsing in the server was presented as 2 stages Erlangian distribution.

4 Conclusions

In this book, we examined different evolutional processes in circuit-switched and packet-based networks, as well as converged network services modeling, performance evaluation, network and services parameters investtigation for the development of the architectures for new services creation and into this converged network implementation.

We have started the main part of our book with the network intelligence evolution process presentation. After the study of a lot of reference materials from different sources (e.g., books, conference and journal articles, companies white papers and technical reports, different industrial news and reports, international bodies and consortia standards, projects and proposals, and so forth), we have seen that circuit-switched (e.g., telephone) networks, for the last 40 years, went through the different evolutional processes. So, we have structurized our original work, and put into some framework, to present it to the readers as a composite evolutional process of telecommunication networks.

Briefly, this process was developing as follows. It started from the simplest conceptual telephone networks, which were the result of years of evolution, with little thought about future technologies. Based almost on analog equipment, telephone networks were not able to support such services as data and video. Many individual technology service providers began popping up providing packet-based networks and data communication services the telephone companies were just not equipped to provide.

Since international bodies began investigating alternative technologies for providing telephone services to the mass, such as mobile, the need for an all-digital network became apparent. Thus, arose the beginnings of an all-digital network with intelligence. The international telecommunication union (ITU) commissioned international telegraph and telephone consultative committee (CCITT) to study the possibility of an all-digital network. The result was a series of standards known now as signaling system no.7 (SS7). These standards have paved the way for the intelligent network (IN), and, with it, a variety of services.

The next step was the distribution of intelligence in telecommunication networks, which began as nothing more than segmentation of responsibilities. The foundations of that segmentation were established according to the trend of moving telecommunication solutions toward more diverse computing platforms and away from monolithic settings. With movement and diversity has come the ability to integrate new solutions into the overall base system with greater speed and efficiency. Ultimately, the base system has transformed to become a part of a larger set of integrated components, each with different levels of responsibility and contribution to the intentions of that evolved solution.

Distributed network intelligence was grounded in the OSI model. It is within this model that the necessary relationships between multiple hosts of the distributed network were established. The OSI reference model defined a partitioning of network operability into seven layers, where one or more protocols implement the functionality assigned to a given layer.

The convergence between fixed and mobile networks that until recently were two separated worlds was next logical step of networks and services intelligence evolution towards single, open platform. The idea to run one single network instead of two separate, one for fixed services and one for mobile services, allows network operators to save money in terms of infrastructure and operational costs. The ability to provide integrated fixed and mobile services offered opportunities both for the provision of promising services and for a recognisable differentiation from competitors. From customers' point of view fixed/mobile convergence has given an integrated service package, whereby the customer was offered both types of services, using one terminal and possibly one number, and received one bill. It has met customers' expectations in terms of independence from terminals and technologies.

And, finally, converged IN, as a single platform for voice, multimedia, and data, could offer these services regardless of the network access and transport technologies used by the carrier and its customers. The unified service content and call control model has allowed to establish hierarchical network intelligence. And, this hierarchy has separated intelligent functions in converged IN.

The converged IN together with its unified service content and call control model was a final phase of the network intelligence evolution. But, it has been evolution process of pure telecommunication networks; in order new services development and support.

In last several years, telecommunication networks have been migrating toward packet-based technologies. In Chapter 3, we examined different integration aspects of IN and IP networks convergence. We have given comprehensive survey of all standardization and research efforts in this direction. First, we have stressed on the objectives, enablers, network configuration, features requirements, and some other open issues of convergence process. Next, we have performed analysis of work of international and industry bodies for the development of standards and architectures for converged networks.

Call control or signaling in converged networks is a very specific. In Internet telephony, the call control functions of a traditional circuit switch are replaced by a device referred to, as a call agent, a SIP server, a H.323 gatekeeper or a feature server. Therefore, we have presented and made analysis and comparison of different call models and interworking between them for transparent service access from various underlying networks. We described also service control functionalities in terms of componentware model (e.g., service platform). The service control consists of two components: call control and messaging control. The call control offers a common call control interface to all services whereas the me-ssage control component provides an abstract interface to services in the service control layer.

Then, we have made the characterization of services that can be recognized as a converged. These services we have divided into two groups: IN service requests towards IP networks, and IN service requests from IP networks. According to that division, we explained in details and made originnal illustrations of signaling messages flows for every service scenario.

And, finally, two protocols PINT (for IN service requests from IP networks) and SPIRITS (for IN service requests towards IP networks) were presented. These protocols were created for converged services support. They are working on the application layer of OSI model (e.g., above SIP protocol). PINT and SPIRITS protocols are mirror images of the integration process. And, we have made some framework and proposed to develop them as a single, joined protocol in converged environment.

Besides, we have performed an analysis and modeling of pure IN and converged services. The analysis and modeling of every network signaling traffic involves statistical mathematics. The buffers in any communication network form queues of requests waiting to be served. It is reasonnable to expect that the amount of data in a queue will have some effect on how long it will take before the data will be served. It is also reasonable to expect that the occupancy of the queue will be determined by a number of factors such as the arrival and departure rates of the data. Therefore, we applied queueing theory methods for the investigation of these factors to determine and calculate IN performance parameters. We defined traffic parameters that needed to be analyzed and investigated. Also, we introduced and explained modeling techniques for the in-vestigation of signaling messages flows in telecommunication networks.

We proposed finite source queueing model for IN signaling flows modeling, and have motivated their usage in this network. We introduced and examined its general properties, and performance parameters to be measured are analyzed. Also, the model shortcomings were given.

The additional load generated by new IN services may lead to a performance degradation that can spread beyond the IN environment, which, in turn, affects not only the quality of the new IN services, but also the services already offered. In our work, the modeling approach was based on the construction of model for the various components of the IN architecture leading to a multiple-chain queuing network system. The analysis was conducted using hierarchical decomposition techniques, allowing a detailed consideration of the signaling network protocol.

Next, we examined, and applied teletraffic theory techniques for converged network services interworking. Based on different reference sources, we proposed and illustrated in details five basic scenarios for services interworking in converged network. Also, we analyzed and evaluated differrent signaling protocols for service scenarios. For every scenario (e.g., signaling messages flows) we applied some queueing system or systems. We evaluated and motivated usage of every queueing system into every scenario implementation. Also, we provided numerical results of important parameters calculations (e.g., time in system, waiting time, server busy period, number of users in system, number of users waiting in queue, and so on), and demonstrated the calculations results in numerous diagrams.

5 References

[3Com00] Technical Paper, "Voice Service Interworking for PSTN and IP Networks", 3ComCorporation, 2000.

[3Com05] White paper, "Taking the guesswork out of deploying IP Telephony", 3Com Corporation, 503147-003 11/05, 2005.

[Ach00] R.A. Achterberg, and H.E. Hanrahan, "Framework for migrating technology in advanced service provision networks", Proceedings of the IEEE Intelligent Networks Workshop (IN2000), Cape Town, South Africa, May 7-11, 2000, pp.130-140.

[Ach02] R.A. Achterberg and H.E. Hanrahan, "Temporal Conceptual Model for Migrating Technology within Telecommunications Systems", Proceedings of the South African Telecommunications Networks and Applications Conference, September 2002.

[Ack97] R. Ackeley, A. Elvidge, T. Ingham, and J. Shepherdson, "Network Intelligence – Performance by Design", IEICE Transactions on Communication, vol. E80-B, no.2, February 1997, pp.219-229.

[Agb07] F. Agboma, A. Liotta, "Managing the user's quality of experience", Proceedings of the 2nd Business-Driven IT Management (BDIM 2007) 21-22 May 2007, Munich, Germany, pp. 98-99.

[All90] A.O. Allen, "Probability, statistics, and queueing theory: with computer science applications" - Second edition, Academic Press, 1990, ISBN 0-12-051051-0.

[And01] A. Andreetto, C.A. Licciardi, P. Falcarin, "Service opportunities for Next Generation Networks", Proceeding of the Eurescom Summit conference - 3G Technologies and Applications, Heidelberg, Germany, November 2001.

[Baf96] M. Bafutto, M. Schopp, "Network Performance and Capacity Figures of Intelligent Networks based on the ITU-TS IN Capability Set-1", Proceedings of the International Workshop on Advanced Intelligent Networks (AIN'96), Passau, Germany, 1996, pp. 15-29.

[Bak00] J.-L. Bakker, J.R. McGoogan, W.F. Opdyke, F. Panken, "Rapid development and delivery of converged services using APIs", Bell Labs Technical Journal vol.5, no.3, 2000, pp. 12-29.

[Bak04] Baken, N., van Boven, E., den Hartog, F., Hekmat, R.: A Four-Tiered Hierarchy in a Converged Fixed-Mobile Architecture, Enabling Personal Networks. The Journal of the Communications Network 3, 2004, pp. 98–104.

[Bas08] R. Basole, W. Rouse, "Complexity of Service Value Networks: Conceptualization and Empirical Investigation", IBM Systems Journal, Vol. 47, Issue 1, Special Issue on Service Science, 2008, pp. 53-70.

[Ber00] M. Bernardi, J. Nuijten, "Final report on number portability for mobile networks", prepared by ETO for ECTRA, 4 April 2000.

[Ber92] D. Bertsekas, R. Gallager, "Data Networks" – Second edition, Prentice-Hall, 1992, ISBN 0-13-200916-1.

[Bha00] R. R. Bhat and R. Gupta, "JAIN Protocol APIs", IEEE Communications Magazine, vol.38, no.1, January 2000, pp.100-107.

[Bhu07] S. Bhulai, R.D. van der Mei, and M. Wu, "Design and optimization of streaming content delivery networks", Proceedings of IEEE GlobeCom, Washington DC, November 2007.

[Bla00] V. Blavette, G. Canal, U. Herzog, C. A. Licciardi, S. Tuffin, "EURESCOM P909 contribution to PINT and SPIRITS Interaction between Internet and PSTN to request services from one domain to the other ", IETF Internet draft, <draft-canal-p909-pint-spirits-00.txt>, March 2000.

[Bou08] C. Boulton, T. Melanchuk, S. McGlashan, "A Basic Interactive Voice Response (IVR) Control Package for the Media Control Channel Framework", draft-boultonivr-control-package-06, February 2008.

[Bre00] R. Brennan, B. Jennings, C. McAedle, and T. Curran, "Evolutionary Trends in Intelligent Networks", IEEE Communications Magazine, Vol.38, No.6, June 2000, pp.86-93.

[Bri05] B. Briscoe, S. Rudkin, "Commercial models for IP Quality of Service Interconnect", BT Technical Journal 23(2), 2005.

[Bro85] I.N. Bronshtein, K.A. Semendyayev, "Handbook of Mathematics", ISBN 3-540-62130-x, 1985.

[Bul00] J. Buller, "A proposal for the provisioning of Call Completion Internet Busy using PINT and SPIRITS", IETF Internet draft, <draft-buller-spirits-ccib-00.txt>, December 2000.

[Car95] B. E. Carne, "Telecommunications Primer: Signals, Building Blocks and Networks", Prentice Hall, 1995, ISBN 0-13-490426-5.

[Cha00] F. Chatzipapadopoulosa, G. De Zenb, T. Magedanz, I.S. Venierisa, F. Zizzab, "Harmonised Internet and PSTN service provisioning", Computer Communications, vol.23, 2000, pp. 731–739.

[Chan00] D. Chantrain, K. Handekyn, H. Vanderstraeten, "The soft terminal: extending service intelligence from the network to the terminal", Alcatel Telecommunications Review - 2nd Quarter 2000, pp.135-141.

[Choi00] H.-O. Choi, Y.-J. Kim, D. Han, and S. An, "Dynamic Queue Management Mechanism for Enhancing Call Completion Rate in Wired/Wireless Intelligent Networks", IEICE Transactions on Communications, vol. E83–B, no.6 June 2000, pp. 1342-1354.

[Coo90] R.B. Cooper, "Introduction to Queueing Theory – 3rd edition", ISBN 0-941893-03-0, CEEPress Books, 1990.

[Cri01] J.C. Crimi, "Next Generation (NGN) Services", White paper, Telcordia Technologies, 2001.

[Dav05] A. Davy, B. Jennings, "Coordinating Adaptation of Composite Services, in New Issues on Coordination and Adaptation Techniques", 2nd International Workshop on Coordination and Adaptation Techniques for Software Entities (WCAT05), Glasgow, Scotland, 2005.

[Dav06] A. Davy, D.Botvich, B.Jennings, "An Efficient Process for Estimation of Network Demand for QoS- Aware IP Network Planning", 6th IEEE Workshop in IP Operations and Management, LNCS Autonomic Principles of IP Operations and Management, vol. 4268/2006, 2006, pp. 120-131.

[Dav94] N.J. Davies, "Performance and Scalability of Parallel Systems", PhD thesis, University of Bristol, 1994.

[Den03] B.A. Denison, C.W. Hoopmann, D.P. Mongeau, E.M. Shepherd, P. Wu, "Intelligent maintenance and management of Service IntelligentTM network architectures", Bell Labs Technical Journal, vol.7, no.4, 2003, pp. 171-185.

[Dia01] Y. Diao, J.L. Hellerstein, S. Parekh, "Stochastic Modeling of Lotus Notes with a Queueing Model", Proceedings of the CMG 2001 International Conference, Anaheim, CA, December 2001, Computer Measurement Group (2001).

[Dial01] Dialogic Corporation White paper "Next Generation Voice Services", 2001.

[Did99] T. M. Didriksen, L. S. Sørumgård, T. Ø. Molnes, "INexpensive Open Distributed Service Platform", IFIP TC6 WG6.7 Fifth International Conference on Intelligence in Networks (SMARTNET'99), Pathumthani, Thailand, 1999, pp. 107-120.

[Dix96] S. Dixit and S. Elby, "Frame Relay and ATM Interworking" IEEE Communications Magazine, vol.34, no.6, June 1996.

[Duf94] D.E. Duffy, A. McIntosh, and W. Willinger. "Analyzing Telecommunications Traffic Data from Working Common Channel Signaling Subnetworks", Bellcore Technical Report TM-ARH-023638, February 1994.

[ECTF] ECTF – Enterprise Computer Telephony Forum

http://www.ectf.org

[EDSC02] Technical paper, "Unified Communications in a Converged Network", Electronic Data Systems Corporation, February 2002.

[Ejz04] R. Ejzak, C. Florkey, R. Hemmeter, "Network Overload and Congestion: A comparison of ISUP and SIP", Bell Labs Technical Journal Lucent Technologies, Inc. 9(3), 2004, pp. 173–182.

[Est01] G.H. Estes, "NGN: Preparing for Tomorrow's Services", Alcatel Telecommunications Review, 2nd Quarter 2001, pp. 82-84.

[ETSI99] ETSI Technical report, "Telecommunications and Internet Protocol Harmonization Over Network (TIPHON); Description of technical issues", – TR 101 300 v2.0.1, May 1999. http://www.etsi.org

[EUR1109] Eurescom Project P1109, "Next Generation Networks: The services offering standpoint", 2001-2002.

[EUR909] EUROSCOM Project P909-GI, "Enabling Technologies for IN Evolution and IN-Internet Integration", Deliverable 2, "Architecture and Service Scenarios for Internet-IN convergence", Annex 3: Reference Architecture, 2000.

[Fal03] P. Falcarin, C.A. Licciardi, Analysis of NGN service creation technologies. In IEC (International Engineering Consortium) "Annual Review of Communications", vol.56, June 2003.

[Fat06] H. Fathi, S. Chakraborty, R. Prasad, "On SIP Session Setup Delay for VoIP Services Over Correlated Fading Channels", IEEE Trans. on Vehicular Technology 55(1), 2006, pp. 286–295.

[Fay00] I. Faynberg, H. Lu, M. Weissman, and L. Slutsman, "Toward Definition of the Protocol for PSTN-initiated Services Supported by PSTN/Internet Interworking", IETF Internet draft, <draft-ietf-spirits-protocol-00.txt>, March 2000.

[Fay00a] I. Faynberg, H. Lu, and L. Slutsman, "IN- and PINT-related Requirements for SPIRITS Protocol", IETF Internet draft, <draft-faynberg-spirits-inpintreqs-00.txt>, July 2000.

[Fay01] I. Faynberg, H. Lu, M. Weissman, and L. Slutsman, "The SPIRITS Architecture", IETF Internet draft, <draft-slutsman-spirits-architecture-01.txt>, 2001.

[Fow96] H. J. Fowler and J.W. Murphy, "Network Management Considerations for Interworking ATM Networks with Non-ATM Services", IEEE Communications Magazine, vol.34, no.6, June 1996.

[Fre02] O. Frelot, J. Taeymans, G. Bonnet, "Introduction to next generation applications", Alcatel Telecommunications Review, 2nd Quarter 2002, pp.121-123.

[Fu05] X. Fu, "NSIS: a new extensible IP signaling protocol suite", IEEE Communications Magazine, 43(10), 2005, pp.133–141.

[Gba98] C. Gbaguidi, J. P. Hubaux, G. Pacifici, and A. N. Tantawi. "An Architecture for the Integration of Internet and Telecommunication Services", EPFL Technical Report SSC/1998/025, 1998. http://sscwww.epfl.ch

[Gur00] V. Gurbani, "Accessing IN services from SIP networks", IETF Internet draft, <draft-gurbani-iptel-sip-to-in-02.txt>, May 5, 2000.

[Hac98] A. Hac, and L. Gao, "Analysis of congestion control mechanisms in an Intelligent Network", International Journal of Network Management, vol. 8, 1998, pp. 18-41.

[Ham99] M. Hamdi, O. Verscheure, I. Dalgic, J.-P. Hubaux, P. Wang, "Voice Service Interworking for PSTN and IP Networks", IEEE Communications Magazine, vol.35, no.5, May 1999, pp. 104-111.

[Han97] M. Handley, H. Schulzrinne, and E. Schooler, "SIP: Session Initiation Protocol", Internet Draft, July 1997.

[Han99] H.E. Hanrahan, "Evolution of Standards for Advanced Telecommunications Services and Network Management", Proceedings of the 2nd Annual South African Telecommunications Networks Applications Conference, Durban, September 1999, pp. 12-17.

[Has80] O. Hashida, T. Ueda, M. Yoshida, and Y. Murao, "Queueing Tables", The Electrical Communication Laboratories Nippon Telegraph and Telephone Public Corporation, Tokyo, Japan, 1980.

[Hig98] G.N. Higginbottom, "Performance evaluation of Communication Networks", Artech House, 1998, ISBN 0-89006-870-4.

[Hoo83] M.H. van Hoorn, "Algorithms and Approximations for Queueing Systems", PhD thesis, Vrije Universiteit te Amsterdam, Mathematical Center, Amsterdam, 1983.

[Hub99] J. P. Hubaux, C. Gbaguidi, S. Koppenhoefer, J. Y. Le Boudec, "The Impact of the Internet on Telecommunication Architectures", Computer Networks and ISDN Systems, Special Issue on Internet Telephony, vol.31, no.3, February 1999, pp. 257-73.

[IBM05] Research paper, "Services over IP - Delivering new value through next-generation networks", IBM Institute for Business Value, G510-6234-00, 2005.

[IEC-DNI] IEC, Web Proforum Tutorials, "Distributed Network Intelligence", The International Engineering Consortium, 1999. http://www.iec.org

[IEC-IN/IP] IEC, Web ProForum Tutorials, "Internet Protocol (IP)/Intelligent Network (IN) Integration", The International Engineering Consortium, 1999. http://www.iec.org

[IEC-IPTC] IEC, Web ProForum Tutorials, "Internet Protocol (IP) – Telephony Clearinghouses", The International Engineering Consortium, 1999. http://www.iec.org

[IEC-SS7] IEC, Web Proforum Tutorials, "Signaling System (SS7) Gateway Solution for Internet Access", The International Engineering Consortium, 1999. http://www.iec.org

[IEC-UM] IEC, Web ProForum Tutorials, "Unified Messaging", The International Engineering Consortium, 2000. http://www.iec.org

[IEC-VDC] IEC, Web ProForum Tutorials, "Voice-Data Consolidation", The International Engineering Consortium, 2000. http://www.iec.org

[Iss05] S. Issaeva, "Fixed-Mobile Convergence: Creating value with successful business models," Pyramid Research. June 2005.

http://www.pyramidresearch.com/

[ITU-E1] ITU-T Recommendation E.164, "The International Public Telecommunication Numbering Plan", 1997.

[ITU-H22] ITU-T Recommendation H.225.0, "Call Signaling Protocols and Media Stream Packetization for Packet-Based Multimedia Communication Systems", 1998.

[ITU-H23] ITU-T Recommendation H.235, "Security and Encryption for H-Series (H.323- and other H.245-Based) Multimedia Terminals", 1998.

[ITU-H24] ITU-T Recommendation H.245, "Control Protocol for Multimedia Communication", 1998.

[ITU-H3] ITU-T, Study Group-11, Study Group-16, Recommendation H.323, "Packet-based multimedia communications systems", 1998.

[ITU-H4] ITU-T Recommendation H.450.1, "Generic Functional Protocol for the Support of Supplementary Services in H.323", (1998).

[ITU-Q12] ITU-T Q.1200 Recommendation Series for Intelligent Network Architectures.

[ITU-Q121] ITU-T Recommendation Q.1211, "Introduction to Intelligent Network Capability Set 1", 1993.

[ITU-Q7] CCITT, Blue Book, Volume IV, Fascicles VI.7-VI.9, "Specifications of Signaling System No.7", Recommendations Q.700-Q.795, International Telecommunication Union, Geneva, 1989.

[ITU-Q9] CCITT, Blue Book, Volume IV, Fascicles VI.10-VI.11, "Digital Subscriber Signaling System No.1 (DSS1)", Recommendations Q.920-Q.940, International Telecommunication Union, Geneva, 1989.

[Ive01] V.B. Iversen, "Fundamentals of teletraffic engineering", Internet book, 2001. www.tele.dtu.dk/teletraffic

[Ive05] V. Iversen, S. Stepanov, "The unified approach for teletraffic models to convert recursions for global state probabilities into stable form", ITC19. In: Proceedings 19th International Teletraffic Congress, Beijing, China, August 29- September 2, 2005, pp. 1559–1570.

[Jab92] B. Jabbari, "Intelligent network concepts in mobile communications", IEEE Communications Magazine, Vol. 30, No. 2, February 1992, pp. 64-69.

[JAIN00] JAINTM White Paper, "JAINTM: Integrated Network APIs for the Java TM Platform", September 2000.

http://java.sun.com/products/jain/

[Jan00] J. Janssen, D. De Vleeschauwer, G. H. Petit, "Delay and Distortion Bounds for Packetized Voice Calls of Traditional PSTN Quality", Proceedings of the 1st IP-Telephony Workshop (IPTel'2000), Berlin, Germany, April 12-13, 2000, pp. 105-122.

[Jha04] U. S. Jha, "Adaptive Communication Services Key to Global Interoperability", Wireless Personal Communications: An International Journal , Volume 29 Issue 1-2 April 2004.

[Jha06] U. S. Jha, "Convergence of Communication, Infotainment, and Multimedia Applications Require Paradigm Shift in Processor Development and Reuse Pattern", Wireless Personal Communications: An International Journal , Volume 37 Issue 3-4 May 2006.

[Kap00] S.Kapur, R.Vij, "Approach for Services in Converged Networks", in IP Telecom Services Workshop 2000, September 11, 2000, GA, USA.

[Kihl97] M. Kihl, and M. Rumsewicz, "Analysis of overload control strategies in combined SSP-SCPs in the Intelligent Network", Proceedings of the 15th International Teletraffic Congress (ITC-15), Washington D.C., US, 1997, pp.1209-1218.

[Kle75] L. Kleinrock, “Queueing Systems. Volume I: Computer Theory”, ISBN 0-471-49110-1, Wiley-Interscience publication, 1975.

[Kle86a] L. Kleinrock, “Queueing Systems. Volume II: Computer Applications”, Wiley-Interscience publication, 1986, ISBN 0-471-49111-X (v.2).

[Kle86b] L. Kleinrock, and R. Gail, “Solution Manual for Queueing Systems. Volume II: Computer Applications”, Technology Transfer Institute, 1986, ISBN 0-942948-01-7.

[Koc02] K.F. Kocan, W.A. Montgomery, S.A. Siegel, R.J. Thornberry Jr., G.J. Zenner, “Service creation for next-generation networks”, Bell Labs Technical Journal, vol.7, no.1, 2002, pp. 63-79.

[Kol98] G. T. Kolyvas, S.E. Polykalas, I.S. Venieris, “Performance evaluation of integrated IN/ISDN signaling platforms”, Computer Communications vol.21, 1998, pp.606-623.

[Kos98] T.J. Kostas, M.S. Borella, I. Sidhu, G.M. Schuster, J. Grabiec, and J. Mahler, “Real-Time Voice over Packet-Switched Networks”, IEEE Network, vol.12, no.1, Jan–Feb 1998.

[Krä76] W. Krämer and M. Lagenbach-Belz, “Approximate Formulae for the Delay in the Queueing System GI/G/1”, Proceedings of the 8th International Teletraffic Congress (ITC-8), Melbourne, Australia, 1976, pp.235-1 – 235-8.

[Kro01] B. Krogfoss, J. Pirot, “Next generation networks: enablers for new business models”, Alcatel Telecommunication Review – 2nd Quarter 2001, pp. 91-96.

[Kry01] N. Kryvinska, H.R. van As, “A Converged Intelligent Network Concept – Joined Future for Telecommunications and the Internet”, 7th International Conference on Intelligence in next generation Networks (ICIN 2001), Bordeaux, France; 2001, pp. 74 – 79.

[Kry02] N. Kryvinska, H. R. van As, S. Brusilovskiy, “Packet Intelligent Networks based on a Potential Signaling System No.8 Targeting towards the Next Generation Business Model”, St. Petersburg Regional International Teletraffic Seminar 2002, St. Petersburg, Russia, 2002, pp. 12 – 22.

[Kry02] N. Kryvinska, H. R. van As, "Queuing System Models for Performance Analysis of Intelligent Networks", London Communications Symposium 2002, London, UK; 09.09.2002 - 10.09.2002; in: "Proceedings of the London Communications Symposium 2002", LCS Proceedings, 2002, pp. 169-172.

[Kry02b] N. Kryvinska, H. R. van As, "From Call Control towards Service Control in Next Generation Networks", Proceedings of the International Conference on Software, Telecommunications and Computer Networks (SoftCOM 2002), Split, Croatia; 08.10.2002; in: "Workshop on Contemporary Communications", (2002), pp. 29 – 33.

[Kry03a] N. Kryvinska, H.M. Nguyen, "Large Intelligent Network Modeling Using M/M/2/K/K System", IEEE 9th Asia Pacific Conference on Communications (APCC2003) in conjunction with 6th Malaysia International Conference on Communications (MICC), 21-24 September, Penang, Malaysia, 2003, pp. 1007-1011.

[Kry03b] N. Kryvinska, S. Lepaja, H.M. Nguyen, "Service and Personal Mobility in Next Generation Networks", The Fifth IEEE International Conference on Mobile and Wireless Communications Networks (MWCN 2003), 27-29 October, Singapore, 2003.

[Kry03c] S. Lepaja, A. Lila, N. Kryvinska, H.M. Nguyen, "A Framework for End-to-End QoS Provisioning in Mobile Internet Environment", The Fifth IEEE International Conference on Mobile and Wireless Communications Networks (MWCN 2003), 27-29 October, Singapore, 2003, pp. 86-89.

[Kry04] N.Kryvinska, "Intelligent Network Analysis by Closed Queueing Models", Kluwer/Springer Journal "Telecommunication Systems", Vol. 27, No. 1, 2004, pp. 85-98.

[Kry06a] N. Kryvinska, "M/GX/1 Queuing System for the Modeling of Real-Time Services over IP Network", Fifth International Symposium on Mathematical Modelling, 5th MATHMOD, 7-10 February, Vienna, Austria, 2006, pp. DMP: 8-1 ÷ 8-10.

[Kry06b] N. Kryvinska, Do van Thanh, "IN-Managed SIP Network: A Case Study in Network Convergence", The IEEE 20th International Conference on Advanced Information Networking and Applications (AINA 2006), 18-20 April, Vienna, Austria, 2006, vol.1, pp. 727-732.

[Kry06c] N. Kryvinska, Do van Thanh, "Enabling Voice Services in Hybrid/Cross Network Framework", Proceedings of the London Communications Symposium 2006, London, UK, 14th-15th September 2006, pp. 57-60.

[Kry07a] N. Kryvinska, P. Zinterhof, Do van Thanh, "New-emerging Service-support Model for Converged Multi-Service Network and its Practical Validation", The IEEE First International Conference on Complex, Intelligent and Software Intensive Systems (CISIS-2007) in conjunction with ARES 2007, 10-13 April, Vienna, Austria, 2007, pp. 100-107.

[Kry07b] N. Kryvinska, P. Zinterhof, Do van Thanh, "An Analytical Approach to the Efficient Real-time Events/Services Handling in Converged Network Environment", The First International Conference on Network-Based Information Systems (NBiS2007), in conjunction with the 18-th International Conference DEXA 2007, September 3 - 4, 2007, Regensburg, Germany, Springer LNCS-4658, pp. 308-316.

[Kry08] N. Kryvinska, "An Analytical Approach for the Modeling of Real-time Services over IP Network", Elsevier Transactions of IMACS, Journal "Mathematics and Computers in Simulation" (MATCOM), Vol. 79 (2008), pp. 980-990.

[Kry08a] N. Kryvinska, Do van Thanh, "Integrated Management Architecture for the Seamless Services Provisioning in Converged Network", American Scientific Publishers (ASP), the Journal of Ubiquitous Computing and Intelligence (JUCI), Vol. 2, No. 2, 2008, pp. 1–7.

[Kühn94] P.J. Kühn, C.D. Pack, R.A.Skoog, "Common Channel Signaling Networks: Past, Present, Future", IEEE Journal on Selected Areas in Communications, vol.12, no.3, April 1994, pp. 383-393.

[Lao05] L. Lao, J.Cui, M. Gerla, "Multicast Service Overlay Design", In: Proceedings of SPECTS 05, Philadelphia, PA, 2005.

[Lee97] Y. Lee, J.S. Song, "Overload control of SCP in Intelligent Network with Priority", IEICE Transactions Communications, Vol. E80-B, No. 11, November 1997, pp. 1753-1755.

[Len01] G. Lenahan "Next Generation Networks – A Practical View of Network Evolution", White paper, (Bellcore), Telcordia Technologies, Inc., 1998-2001.

[Lic01] C. A. Licciardi, G. Canal, A. Andreetto, P. Lago, "An architecture for IN-internet hybrid services", Elsevier Computer Networks, vol. 35, 2001, pp. 537-549.

[Lic02] C. A. Licciardi, P. Falcarin, "Next Generation Networks: The services offering standpoint. In Comprehensive Report on IP services, Special Issue of the International Engineering Consortium, October 2002.

[Lio08] A. Liotta, D. Tyrode-Goilo, A. Oredope, "Open source mobile VPNs over converged all-IP networks", Journal of Network and Systems Management, 16(2), 2008, 163-181.

[Lip77] L. Lipsky, J.D. Church, "Applications of a Queueing Network Model for a Computer System" ACM Computing Surveys, vol. 9, No. 3, September 1977, pp.205-221.

[Lul06] M. Lulling, J. Vaughan, "A Simulation-based Comparative Evaluation of Transport Protocols for SIP", Computer Communications 29(4), 2006, pp. 525–537.

[Mas01] E. Masuda, T. Mishima, N. Takaya, K. Nakai, and M. Hirano, "A Large-Capacity Service Control Node Architecture Using Multicasting Access to Decentralized Databases in the Advanced Intelligent Network", IEICE Transactions on Communication, vol. E84-B, No.10, October 2001, pp.2768-2780.

[Mee06] S. van der Meer, A. Davy, S.Davy, R. Carroll, B. Jennings, J. Strassner, "Autonomic Networking: Prototype Implementation of the Policy Continuum", 1st IEEE International Workshop on Broadband Convergence Networks (BcN 2006), IEEE, New York, 2006, pp. 1-10.

[Mei05] R. van der Mei, H. Meeuwissen, and F. Phillipson, "Realizing end-to-end Quality of Service for voice-over-IP in a heterogeneous multi-domain network environment", Proceedings 19th International Teletraffic Congress, ITC-19, Beijing, September 2005, pp. 1109-1120.

[Mei07] R.D. van der Mei, "On a unifying theory on polling models in heavy traffic", In: Managing Traffic Performance in Converged Networks - the Interplay between Convergent and Divergent Forces, ITC2007, LNCS-4516, pp.556-567.

[Min00] R. Minerva, C. Moiso, "Will the "Circuits to Packets" Revolution pave the way to the "Protocols to APIs" Revolution" in Proc. 6th Int. Conf. On Intelligence in Networks, Bordeaux, France, January 17-20, 2000, pp. 11-16.

[Mit07] D. Mitra, "Topics in Networking Research", Managing Traffic Performance in Converged Networks, 20th International Teletraffic Congress, ITC20 2007, LNCS-4516, Ottawa, Canada, June 17-21, 2007.

[Mit98] I. Mitrani, "Probabilistic Modeling", Cambridge University Press, 1998, ISBN 0-521-58511-2.

[Miz97] O. Mizuno, A. Shibata, T. Okamoto, and Y. Niitsu, "Models for service management programmability in advanced Intelligent Network", IEICE Transactions Communications, Vol. E80-B, No. 6, June 1997, pp. 915-921.

[Mod00] A. R. Modarressi, S. Mohan, "Control and Management in Next-Generation Networks: Challenges and Opportunities", IEEE Communications Magazine, vol.38, no.10, October 2000, pp. 94-102.

[Moh02] S. Mohapi and H.E. Hanrahan, "Value-based service differentiation for distributed NGN service platforms", Proceedings of the South African Telecommunications Networks and Applications Conference, September 2002, ISBN 0-620-29432-9.

[Mol88] M. K. Molloy "Fundamentals of performance modeling", ISBN 0-02-381910-3, 1988.

[Mor00] L. Moretti, C. Moiso, "Fixed, mobile, and data convergence", Proceedings of the IEEE Intelligent Networks Workshop (IN2000), Cape Town, South Africa, May 7-11, 2000, pp. 401-411.

[Mur04] E. Muraskin, "Report: Hosted IP Voice Will Storm the Enterprise." eWeek. September 29, 2004.

http://www.eweek.com/article2/0,1759,1662501,00.asp

[Noe07] E. Noel and C. Johnson, "Initial Simulation Results That Analyze SIP Based VoIP Networks Under Overload", Managing Traffic Performance in Converged Networks, 20th International Teletraffic Congress, ITC20 2007, LNCS-4516, Ottawa, Canada, June 17-21, 2007.

[Oec07] S. Oechsner and P. Tran-Gia, "Performance Evaluation of a Reliable Content Mediation Platform in the Emerging Future Internet", Managing Traffic Performance in Converged Networks, 20th International Teletraffic Congress, ITC20 2007, LNCS-4516, Ottawa, Canada, June 17-21, 2007.

[Ore07] A. Oredope, A. Liotta, "A performance based evaluation of SIP signalling across converged networks", Proceedings World Congress on Engineering (WCE 2007), London, UK, July 2-4, 2007, pp. 1518-1523.

[Parlay] Parlay Industry Working Group: CORBA IDL. http://parlay.org

[Pat97] A. Patel, S. O'Connell, "A timestamp model for determining real-time communications in intelligent networks", Computer Communications vol.20, 1997, pp.211-218.

[Pet96] B. Petri and D. Schwerje, "Narrowband ISDN and Broadband ISDN Service and Network Interworking", IEEE Communication Magazine, vol.34, no.6, June 1996.

[Por97] P. Porkka, and K. Raatikainen, "CORBA Access to Telecommunications Databases", In Intelligent Networks and Intelligence in Networks, Gaiti, D. (ed.), Chapman & Hall, September 1997.

[Ram00] R. Ramming, "Intelligent network services for converged voice and data networks", Proceedings of the IEEE Intelligent Networks Workshop (IN2000), Cape Town, South Africa, May 7-11, 2000, pp. 269-276.

[Red95] F. J. Redmill and A.R. Valdar, "SPC Digital Telephone Exchanges", Peter Peregrinus, 1995, ISBN 0-86341-301-3.

[Rin99] J. Rinde, "Telephony in the year 2005", Computer Networks, vol.31, no.3, March 1999, pp.157-168.

[Rus95] T. Russell, "Signaling System #7", McGraw-Hill, 1995, ISBN 0-07-054991-5.

[Ryu06] M. Ryu, "QoS class mapping over heterogeneous networks using Application Service Map", ICNICONSMCL, 2006.

[Ryu08] M. Ryu, Y. Kim, and H. Park, "Systematic QoS Class Mapping Framework over Multiple Heterogeneous Networks", Int. Conference on Next Generation Teletraffic and Wired/Wireless Advanced Networking, LNCS-5174, 2008, pp. 212-221.

[Sch00] H. Schwarze and H. Weik, "Migration from Switched Circuit Networks to Packet Network", Proceedings of the 1st IP-Telephony Workshop (IPTel'2000), Berlin, Germany, April 12-13, 2000, pp. 147-149.

[Schn00] S. Schneiders, "Applications and Services for Voice/Data Convergence", Proceedings of the 1st IP-Telephony Workshop (IP-Tel'2000), Berlin, Germany, April 12-13, 2000, pp. 79-84.

[Schw88] M. Schwartz, "Telecommunication Networks: Protocols, Modeling and Analysis", Addison-Wesley Publishing Company, 1988, ISBN 0-201-16423-X.

[Sid01] G. Sidhu, "High-Availability Considerations for Softswitch-Based Networks", White paper, Cable&Wireless, 2001.

[Sij00] P.G.A. Sijben, J.P.L. Segers, L.F.A. Spergel, J. Kozik, "Building the bridge: Devising an architecture to migrate voice-band calls to packet transport and multimedia services", Bell Labs Technical Journal vol.5, no. 3, 2000, pp. 166-185.

[SPIRITS] The Services in the PSTN/IN Requesting InTernet Services (SPIRITS) IETF Working Group.

http://www.ietf.org/html.charters/spirits-charter.html

[Tai99] J. Taina and K. Raatikainen, "Requirements Analysis of Distribution in Databases for Telecommunications", Databases in Telecommunications; International Workshop, Co-located with VLDB-99 Edinburgh, Scotland, UK, September 6th, 1999, Proceedings, LNCS 1819, Springer, 2000.

[Tan95] M. Tanner, "Practical Queueing Anylysis", IBM McGraw-Hill Series, 1995, ISBN 0-07-709078-0.

[Telc00] Telcordia Technologies White Paper, "Intelligent Network Services: Solutions for Fixed-Mobile Convergence", 2000.

[Telc01] Issue Brief #2, "The Next Generation Networks – How Do Intelligent Networks Fit In", Telcordia Technologies, Inc., 1998-2001. http://www.telcordia.com/newsroom/

[Thö94] J. Thörner, "Intelligent Networks", Artech House, 1994, ISBN 0-89006-706-6.

[TRMS-01] Technical Report MS-CIS-01-31, "Overview of Voice over IP", University of Pennsylvania, February 2001.

[Ueb01] R. Uebele, M. Verhoeyen, "Strategy for migrating voice networks to the next generation architecture", Alcatel Telecommunications Review, 2nd Quarter 2001, pp.85-90.

[Van01] H. Vanderstraeten, "Myriad: an application service-enabling platform", Alcatel Telecommunications Review, 3nd Quarter 2001, pp.192-193.

[Van99] G. Vanecek, N. Mihai, N. Vidovic, and D Vrsalovic, "Enabling Hybrid Services in Emerging Data Networks", IEEE Communications Magazine, vol.37, no.7, July 1999, pp. 102-109.

[Vast] K. S. Vastola, "Performance Modeling and Analysis of Computer Communication Networks", Electrical Computer and Systems Engineering Dept. Rensselaer Polytechnic Institute Troy, NY.

http://poisson.ecse.rpi.edu/~vastola/pslinks/perf/perf.html

[Wal00] P. Wallace, B. Farabet, and V. Bic, "Java in Intelligent Networks", in Proceedings 6th International Conference on Intelligence in Networks, Bordeaux, France, January 17-20, 2000, pp.2-10.

[Wan07] S. Wanke, M. Scharf, S. Kiesel, S.Wahl, "Measurement of the SIP Parsing Performance in the SIP Express Router", In: Pras, A., van Sinderen, M. (eds.) EUNICE 2007, LNCS, vol. 4606, pp. 103–110.

[Wei98] M. Weiss, and J. Hwang, "Circuit Switched Telephony or Itel: Which is Cheaper?", The 26th Telecommunications Policy Research Conference Proceedings, October 3-5, Alexandria, VA (1998).

[Whi75] J.A. White, J.W. Schmidt, G.K. Benett, "Analysis of Queueing Systems", ISBN 0-12-746950-8, Academic Press, 1975.

[Yad98] T. Yada, I. Nakajima, I. Ide, and H. Murakami, "Forecasting Traffic Volumes for Intelligent Telecommunication Services Based on Service Characteristics", IEICE Transaction on Communications, vol.E81-B, no.12, December 1998.

[Yan94] J. Yan and D. MacDonald, "Teletraffic Performance in Intelligent Network Services", Proceeding of the 14th International Teletraffic

Congress (ITC-14), Antibes, Juan-les-Pins, France, June 1994, pp.357-366.

[Zep94] J. Zepf, and G. Rufa, "Congestion and Flow Control in Signaling System No.7 – Impacts of Intelligent Networks and New Services", IEEE Journal on Selected Areas in Communications, vol.12, no. 3, April 1994, pp.501-509.

[Zna97] S. Znaty, J. P. Hubaux, "Telecommunications Services Engineering: Principles, Architectures and Tools", Proceedings of ECOOP`97, Workshop on OO technology for telecommunications Services Engineering, December 1997.

[Zui96] H. Zuidweg, P. Quentin, G. Reyniers, E. Devleeschouwer, B. Quirynen, "A Distributed CORBA-Based IN Architecture", 4th International Conference in Networks, Bordeaux, France, 1996.

[RFC 2848] S. Petrack and L. Conroy, "The PINT Service Protocol: Extensions to SIP and SDP for IP Access to Telephone Call Services", IETF RFC 2848, June 2000. http://www.ietf.org/rfc/rfc2848.txt

Electronic Business

Herausgegeben von Christine Strauss

Band 1 Rudolf Hartjes: Web Accessibility. Techniken und exemplarische Erfolgsmessung. 2009.

Band 2 Natalia Kryvinska: Converged Network Service Architecture. A Platform for Integrated Services Delivery and Interworking. 2010.

Band 3 Marie-Luise Leitner: Business Impacts of Web Accessibility. A Holistic Approach. 2010.

www.peterlang.de